Cosmology 101

COSMOLOGY 101

KRISTINE M. LARSEN

SCIENCE 101

GREENWOOD PRESS
Westport, Connecticut • London

Library of Congress Cataloging-in-Publication Data

Larsen, Kristine M., 1963–
    Cosmology 101 / Kristine M. Larsen.
        p.   cm.—(Science 101, ISSN 1931–3950)
    Includes bibliographical references and index.
    ISBN 978–0–313–33731–4 (alk. paper)
    1. Cosmology—Popular works.   I. Title.
QB982.L37   2007
523.1—dc22   2007000425

British Library Cataloguing in Publication Data is available.

Library of Congress Catalog Card Number: 2007000425
ISBN-13: 978–0–313–33731–4
ISBN-10: 0–313–33731–4
ISSN: 1931–3950

First published in 2007

Greenwood Press, 88 Post Road West, Westport, CT 06881
An imprint of Greenwood Publishing Group, Inc.
www.greenwood.com

Printed in the United States of America

∞™

The paper used in this book complies with the
Permanent Paper Standard issued by the National
Information Standards Organization (Z39.48–1984).

10 9 8 7 6 5 4 3 2 1

Copyright Acknowledgments

The author and publisher gratefully acknowledge permission for use of the
following material:

All figures courtesy of Bridgette Alsbury.

In grateful recognition of all those who have been my teachers.

CONTENTS

Series Foreword

What should you know about science? Because science is so central to life in the twenty-first century, science educators believe that it is essential that *everyone* understand the basic foundations of the most vital and far-reaching scientific disciplines. *Science 101* helps you reach that goal—this series provides readers of all abilities with an accessible summary of the ideas, people, and impacts of major fields of scientific research. The volumes in the series provide readers—whether students new to science or just interested members of the lay public—with the essentials of a science using a minimum of jargon and mathematics. In each volume, more complicated ideas build upon simpler ones, and concepts are discussed in short, concise segments that make them more easily understood. In addition, each volume provides an easy-to-use glossary and an annotated bibliography of the most useful and accessible print and electronic resources that are currently available.

PREFACE

The author of this volume has been charged with a seemingly impossible task—explaining the entire life and make-up of the universe in around 200 pages. Although I have tried my best to live up to the spirit of this assignment, clearly a few details have had to have been left out in the name of brevity. Although these few beginning sentences have been typed with tongue firmly implanted in cheek, the truth is not far away. Cosmologists have really taken upon themselves the study of all aspects of this immense and ancient universe, aided by ever-improving technology, a seemingly limitless source of imagination and ingenuity, and the scientific method. Perhaps those are the truly basic themes of this text. You, the reader, will be my companion on an exhilarating, and admittedly sometimes confusing, journey through space and time. Together we will discover how the human mind, limited by the size of its enclosing skull and its location on the surface of the Earth at this era in time, have sought the answers to the most ancient and fundamental questions—where did the universe come from and where is it going?

To study cosmology in isolation from its history is certainly to ignore the scientific method, and we certainly shall not make that mistake here. We must also never forget the important concept that science is done by scientists, some of whom are more colorful in personality than others. As members of the human race, individual scientists can also make mistakes, can sometimes turn a blind eye toward results that contradict their own research, and can make disparaging comments about the research of others that might not be considered "scholarly." Welcome to the true world of science, warts and all. Some of what you will read here is highly speculative, and a (hopefully very) few items of discussion may be out of date by the time you hold this volume in your hands. If so, consider it yet another valuable lesson in the scientific method, courtesy of this well-meaning and conscientious teacher.

While I have largely circumvented the mathematical nuts and bolts of cosmology in the telling of this tale, it is impossible to avoid numbers altogether. In our voyage through space and time we will necessarily encounter numbers both inconceivably small and astoundingly large. It is for this reason that you will soon be introduced to the mathematical shorthand scientists use to manage numbers of extreme size, known as scientific notation. Otherwise, this book would waste far too many pages devoted to strings of zeros!

This project would not have been possible without the valuable input of Ronald Mallett and Emma Keigh, and would have been far less interesting without Bridgette (Brie) Alsbury's diagrams. You all have my eternal gratitude for all your help.

Now let's start exploring!

INTRODUCTION

WHERE DID IT ALL COME FROM?

Midday, March 29, 2006, the Libya-Egypt border. A cheer slowly erupted as the shadow of the moon raced across the landscape from west to east, enveloping the gathered crowd of 14,000 men, women, and children. With one last, brilliant gasp of light—the diamond ring effect—the sun was swallowed by the dark disk of the new moon, plunging the now-applauding witnesses into an eerie, 360-degree twilight. Astronomers (both professional and amateur), public servants and military men of all ranks, politicians from several nations (including Hosni Mubarek, the president of Egypt), and general devotees of exotic adventures stood shoulder to shoulder as the heavens graced us with a show unlike any other—a total solar eclipse. For that all-too-brief instant in time, divisions of race, religion, language, social class, and political affiliation were utterly irrelevant and forgotten, as we stood in awe on that isolated desert plateau, equal citizens of a vast and fascinating cosmos. A far-too-fleeting four minutes later, the sun reappeared with an equally evanescent diamond ring, signally the end of totality. "Well, Rahu gave back the sun," I kidded with my companions, making reference to the ancient Hindu myth for the cause of solar eclipses. After what I had just witnessed, it was more obvious to me than ever why our ancestors across the globe revered the heavens, worshipping the stars, planets, sun, and moon as gods and goddesses, and generally sought explanations for the forces of nature in a creative and colorful pantheon of demons and deities.

This universal need to explain the world and its workings forms the basis of the uniquely human activity called myth. These traditional, ancient stories were crafted by cultures according to their own interactions with the environment, and sought to explain the beginnings of the world,

the natural patterns of birth and death, and outlined the culturally ac-
ceptable behavior of that particular group of people. These stories have
been passed down largely through oral traditions from generation to
generation, which explains the normal variations seen in most myths.
Among the most fundamental myths are those that seek to explain the
creation of the earth and heavens, a culture's *creation myth*. As Barbara
Sproul (1979, 1) explains, "The most profound human questions are
the ones that give rise to creation myths: Who are we? Why are we here?
What is the purpose of our lives and our deaths? How should we understand our place in the world, in time and space?" Understandably, many
religions include a creation myth among their traditions.

Although there is a rich variety of creation myths from around the
world, there are certain common threads or archetypes which are seen
repeatedly. The central theme is usually the creation of the universe and
its various parts. Some models have a one-time creation, while others
explain the universe as being a giant cycle of creation and destruction.
Some envision the universe as being created from nothingness, or chaos.
The creator can be seen as a single god or goddess, or a group of beings.
The process of this creation can be a celestial war, the death of a sacrificial
god, the separation of earth from sky, or a special song, dance, or series
of spoken words (such as the famous "Let there be light" of the Old
Testament).

For example, in a Chinese myth, a great egg hatched after the passing
of numerous centuries, bringing forth from his lengthy sleep the giant
creature Pan Gu. The top of the shell became the heavens, while the
bottom of the shell became the earth. He kept the two separate using his
ever-growing body. After the heavens became fixed into place, Pan Gu
died. His breath gave rise to the wind, his voice gave birth to thunder, and
his eyes became the sun and moon. According to Cherokee legend, the
world was originally covered with water. Through the action of Lock-
chew, an industrious crawfish, mud was brought up from below the
waters to make land, which was in turn dried by the flapping of the
buzzard Yah-tee's wings. The human race sprang from drops of blood
from T-cho, the Sun goddess. Even fictional cultures can have a creation
myth, such as the elves in *The Silmarillion*, J.R.R. Tolkien's sweeping
epic prequel to *The Lord of the Rings*. After tackling the formation of the
universe, and most importantly Earth itself, most myths then explain the
origin of the sun, moon, and stars, and finally human beings and other
citizens of the world. Some myths continue further, offering a vision
of the end of the world, and the ultimate triumph over evil and/or
the death of the gods, such as the Ragnarok—"Doom of the Gods"—of
Norse myth.

Thanks to meticulous observations of the heavens, both with the unaided eye and with increasingly powerful technology (including the invention of the telescope, photography, and spectroscopy), coupled with mathematical techniques and advances in our understanding of the basic forces of nature (such as gravity), a scientific understanding of the universe, called *cosmology*, has developed over the past two millennia. Derived from the Greek "kosmos" (order), cosmology is the scientific study of the history and structure of the universe. Sometimes considered a special branch of astronomy, it draws from myriad other areas of study, including mathematics, physics, and chemistry. Although cosmology traces its origin to those ancient and poetic explanations offered in creation myths, modern cosmology is indisputably a discipline of science, and follows the *scientific method* in exploring different possible models of the history of the universe.

What Is Science?

Science is a methodology used to explore the natural world and is based on the twin pillars of observation and theory. Countless observations are made of the world around us, both passively collecting information from natural occurrences, as well as actively performing specific experiments in the laboratory which probe nature at a deeper level. Based on those observations, scientists develop *hypotheses*—proposed explanations which can then be tested and further refined (or refuted). As a hypothesis is continually tested, it will become accepted by more scientists if it is found to successfully explain observations and experimental outcomes, and becomes a *theory*. Unfortunately, in the common vernacular, the word "theory" has become synonymous with a guess or a hunch, while a scientific theory is actually neither.

Regardless of specific discipline, scientific theories have several common attributes: they are consistent, both internally and with already observed phenomenon; they are falsifiable (a concept widely explored by philosopher Karl Popper); they make predictions that can be observed with current or future technology; they tend to be relatively simple or elegant (although it might seem the opposite to the casual observer); finally, a theory can never be proven absolutely true, therefore there is no end to scientific endeavor. A true scientific theory is always open to being disproved, and the burden of proof is continually placed on the scientist. It is vitally important that observations and experiments, as well as theoretical calculations, be repeatable by various independent groups of scientists, to minimize error and bias. This is commonly called "peer review." It is this peer review that assures the fundamental integrity of science, if not most scientists. In the words of astronomer Vera Rubin

(1997, 219), "Science is competitive, aggressive, demanding. It is also imaginative, inspiring, uplifting." At its very core, the process of science is based on gathering knowledge and revising our understanding based on that knowledge. There is no such thing as "faith" in science, no "sacred truths"—everything is open to revision. No matter how successful a theory may be, it is always possible that new technology or theoretical investigations could spell its demise. The late astrophysicist John Bahcall wistfully remarked that "every time we get slapped down, we can say 'Thank you Mother Nature,' because it means we're about to learn something important" (Lemonick and Nash 1995, 84).

An important distinction needs to be drawn between the scientific method, which is designed to be as free from prejudice as possible, and scientists, who, as human beings, are, unfortunately, sometimes susceptible to personal bias. For example, the big bang theory was disdained by many scientists in the 1940s and 1950s because they felt it was too "theological" (in that it suggested a beginning of the universe rather than an eternal state of being). There can even be "fads" in science, hypotheses which become suddenly popular for a time, only to be abandoned just as quickly. Astronomer Donald Fernie has lamented this herd mentality of scientists, especially his colleagues, whom he compares to "antelope, heads down in tight formation, thundering with firm determination in a particular direction across the plain. At a given signal from the leader we whirl about, and, with equally firm determination, thunder off in a quite different direction, still in tight parallel formation" (Singh 2004, 378–379). Scientists can even become personally attached to their theories, being convinced of the beauty of the idea and reticent to abandon it. One particularly sarcastic comment is that "a competent astrophysicist can reconcile any theory with any new observation," while an "even more cynical colleague extended this claim, asserting that the astrophysicist often need not even be competent" (Rees 2000, 39).

On occasion, the scientific establishment has even marginalized new ideas and observations which contradict the "canon" of the field. An example is the measurement of the expansion rate of the universe made by Wendy Freedman and her research group in the 1990s using the Hubble Space Telescope. When her measurements contradicted then current assumptions about the age of the universe, some astronomers dismissed her work out of hand. Despite the human foibles of scientists, science as a process does prevail, as in the case of Freedman's work, which was subsequently vindicated by further observations. The key is that because the scientific community relies on peer review, claims of either new observations or theories are repeatedly checked by colleagues at other institutions. In this way, claims that have no basis, or outright frauds (such as the announcements of successful human cloning by South Korean scientist Hwang Woo-suk) are eventually revealed. This self-correcting mechanism is one of the key aspects of science.

Science and Religion

In the words of theorist Paul Davies,

No science is more pretentious than physics, for the physicist lays claim to the whole universe as his subject matter.... Physicists, like theologians, are wont to deny that any system is in principle beyond the scope of their subject. (Davies and Brown 1988, 1)

Based on comments such as this, as well as issues in the news, such as debates over stem cell research and the teaching of "Intelligent Design," it is understandable that there is widespread confusion concerning the relationship between science and religion. In general, the relationship between religion and science can be respectful or adversarial. One point of view is that they ask fundamentally different questions and seek different answers. In this model, there is no cause for conflict. Diametrically opposed to this is the viewpoint that there will always be tension between the two fields because they seek to explain the same world through completely different lenses. An example of the first perspective is found in the writings of Tenzin Gyatso (2005, 206), the Fourteenth Dalai Lama, who explains that

Many aspects of reality as well as some key elements of human existence, such as the ability to distinguish between good and evil, spirituality, artistic creativity—some of the things we most value about human beings— inevitably fall outside the scope of the [scientific] method.

Physicist Stephen Hawking agrees: "Love, faith, and morality belong to a different category to physics. You cannot deduce how one should behave from the laws of physics" (1993, 173).

In this vein, paleontologist Stephen Jay Gould (1999, 5) suggested a "principle of respectful noninterference . . . by enunciating the Principle of NOMA, or Non-Overlapping Magisteria." Gould was careful to define the distinct magisteria as follows:

Science covers the empirical realm: what is the universe made of (fact) and why does it work this way (theory). The magisterium of religion extends over questions of ultimate meaning and moral value. These two magisteria do not overlap, nor do they encompass all inquiry (consider, for example, the magisterium of art and the meaning of beauty). To cite the old clichés, science gets the age of rocks, and religion the rock of ages; science studies how the heavens go, religion how to go to heaven. (Gould 1999, 6)

In this way, there is no inherent conflict between science and religion. Abbé Georges Lemaître, one of the founders of the big bang model, similarly wrote that

Once you realize that the Bible does not purport to be a textbook of science, the old controversy between religion and science vanishes. . . . There is no reason to abandon the Bible because we now believe that it took perhaps ten thousand million years to create what we think is the universe. Genesis is simply trying to teach us that one day in seven should be devoted to rest, worship, and reverence—all necessary to salvation. (Guth 1997, 55)

Framing the relationship between science and religion in this way brings attention to a point frequently lost in the ongoing debates, namely that rather than being antireligous, science is unreligious. Likewise, in cosmological discussions, physicists, such as Stephen Hawking (1993, 99), openly admit that there is still room for religion:

And even if there is only one unique set of possible laws, it is only a set of equations. What is it that breathes fire into the equations and makes a universe for them to govern? . . . Although science may solve the problem of how the universe began, it cannot answer the question: Why does the universe bother to exist?

With the respectful relationship of NOMA, however, also comes responsibilities on both sides. Religion has no basis to impose on scientific theory (such as Young Earth Creationism, or "Intelligent Design"), while as Gould (1999, 9–10) warns, "scientists cannot claim higher insight into moral truth from any superior knowledge of the world's empirical constitution." Clearly failure on either side to respect the boundaries of their magisteria leads to conflict. Problems arise between science and religion when the point of view is taken that each one has a lock on explaining the one true reality. For example, a religious group or movement may ignore the scientific method and cling to beliefs about the world which are plainly contradicted by observations (such as the age of Earth). As the Dalai Lama admonishes, "spirituality must be tempered by the insights and discoveries of science. If as spiritual practitioners we ignore the discoveries of science, our practice is also impoverished, as this mind-set can lead to fundamentalism" (Gyatso 2005, 13). On the other side, scientists may claim that the scientific method, rather than being silent on supernatural possibilities, negates the possibility of God. Both points of view are equally counterproductive.

The benefits of a harmonious, integrated relationship between science and religion have been expressed by scientists and religious figures alike. On one side, it could create a bridge between science (and scientists) and the larger human experience. As the Dalai Lama offers,

Since the emergence of modern science, humanity has lived through an engagement between spirituality and science as two important sources of knowledge and well-being. Sometimes the relationship has been a close one—a kind of friendship—while at other times it has been frosty, with many finding the two to be incompatible. . . . science and spirituality have the potential to be closer than ever and to embark upon a collaborative endeavor that has far-reaching potential to help humanity meet the challenges before us. We are all in this together. May each of us, as a member of the human family, respond to the moral obligation to make this collaboration possible. This is my heartfelt plea. (2005, 208–209)

On the other side, religious practitioners have found that an understanding of science fosters an appreciation for the wonders of the universe and can actually deepen their faith. Carl Sagan (1996, 29–30) concurs, writing that

Science is not only compatible with spirituality; it is a profound source of spirituality. When we recognize our place in an immensity of light-years and in the passage of ages, when we grasp the intricacy, beauty, and subtlety of life, then that soaring feeling, that sense of elation and humility combined, is surely spiritual. . . . The notion that science and spirituality are somehow mutually exclusive does a disservice to both.

Cosmology owes its origins to spiritual notions about the natural world; it is fitting that it still peacefully coexists with that aspect of human activity in this modern age.

1

LUMINOUS MATTERS:
LIGHT AND STARS

COSMOLOGY AS A UNIQUE PRACTICE

Cosmologists are at a serious disadvantage compared with other scientists. Biologists can directly experiment on specimens in the laboratory, raising generations of mice, bacteria, corn plants, or fruit flies, carefully monitoring how they interact, age, and evolve. Geologists can sample rocks from around the world, created during different periods in the earth's history, and conduct tests in the lab to determine their age and composition. Chemists can mix together numerous compounds in the laboratory in varied proportions and under different conditions, and repeat the same experiments over and over again. Astronomers in general, and most specifically cosmologists, can only, as your mother warned, "look but don't touch." Limited by what Peebles and Ratra (2003, 560) call this "astronomers' Tantalus principle," we can't gather a sample of stars and bring them into the laboratory to dissect. Neither can we build our own stars in the laboratory and watch them age and die over a year or two (the time of the average funding grant). It would be nice if the universe would repeat the experiment of the big bang before our eyes so we could test our theories more directly, but since that might necessitate our own destruction, perhaps we should humbly accept our passive position as mere observers some 13 billion years after the fact without complaint. The point is that, since cosmologists are largely limited to observing what events the universe deigns to show us, in whatever frequency and style it decides to do that, creative means are required to wring every bit of information out of each light beam which comes our way. This is, of course, one of the reasons why cosmology has exploded as a field over the past century as a direct result of improvements in technology.

The study of cosmology requires being conversant in a new language, that of mathematics. For example, cosmology is a science of very big (and very small) numbers. Scientific notation is a shorthand way of taking care of all those nasty zeroes (thereby cutting down the length of astronomy textbooks). For example, six million is written as 6,000,000 or 6×10^6. On the other hand, six millionths is written as 0.000006, or 6×10^{-6}. In each case, the exponent (or power of ten) counts the number of places the decimal point is moved either to the right or left of its usual position. Most scientists utilize one of two standard sets of metric system units to describe physical properties such as length, mass, and time. These are the so-called mks (meters/kilograms/seconds) and cgs (centimeter/gram/seconds) systems. Astronomers and cosmologists are notorious for augmenting these standard systems with unique units they have contrived in an effort to make their equations "prettier." Two of the most important are the *astronomical unit* or au (defined as the average distance between the Sun and Earth, 93 million miles or 150 million kilometers) and the *parsec*. The definition of a parsec is difficult to understand out of the context of its development, so we will leave that for future discussion. For the moment, know that one parsec is 206,265 au, defined in such a way as to exactly "cancel out" a factor of (you guessed it) 206,265 in an important astronomical equation! The parsec (or pc) is the unit astronomers and cosmologists use to discuss the distances to stars. For larger distances, kiloparsecs (kpc) or thousands of parsecs, and megaparsecs (mpc), millions of parsecs, are routinely used. Astronomers also use the star they understand the best, namely our sun, as a common yardstick against which to measure other stars. For example, a solar mass unit measures the mass of a star relative to the mass of the sun.

But what about the so-called light year commonly used in science fiction? Although there is certainly some affection attached to the term, it is not commonly used by astronomers when talking among each other, as it was somewhat artificially defined, does not simplify equations, and actually leads to confusion when communicating with the general public, who commonly believe it to be a measure of time and not of length. As defined, a light year is the distance a beam of light could travel in one year. Traveling at a speed of 186,000 miles (300,000 km) per second, the corresponding distance amounts to about six trillion (6×10^{12}) miles (9×10^{12} km). One parsec is 3.26 light years. By comparison, the nearest star (the Sun), is 93×10^6 miles away. It takes a beam of sunlight just over 8 min to reach Earth, and nearly six hours to reach Pluto. It takes that same beam of sunlight over 4 years to reach Alpha Centauri, the nearest star system outside of our own. Our intrepid ray of solar radiation would spend over 24,000 years trying to reach the center of our Milky Way Galaxy, and more than 2 million years crossing intergalactic space to reach the Andromeda Galaxy, the nearest large spiral galaxy to our own.

If your head is beginning to ache, don't despair, but be warned that the pain is about to be ratcheted up just a tad bit more. Turn these numbers around for a second and think about what they really mean. If it takes sunlight 8 min to reach Earth, when you look up at the Sun, you are not seeing it as it really is this moment, but how it *was* 8 min ago. If the sun were to really be swallowed by the demon Rahu, we wouldn't know until 8 min after the crime had been committed. Pluto would continue to receive "old" sunlight for six hours before noticing something was amiss, and any Alpha Centaurians would only notice after 4 years' time that a certain yellow star had suddenly disappeared. Any observers in the Andromeda Galaxy would continue to see our sun for over 2 million years after its demise! This means that the news from Earth is always woefully out of date when seen from cosmic distances, and our view of the universe is also past history. The farther out we look into space, the farther back we see in time. This concept is sometimes called look-back time, and is routinely the source of wonder and some bewilderment for poor unsuspecting astronomy students. Such is the nature of cosmology. As we shall see on our journey together through the universe, it is "not only queerer than we suppose, but queerer than we can suppose" (Yulsman 2003, 38).

THE ELECTROMAGNETIC SPECTRUM

Light is usually thought of as a wave produced by oscillating electric and magnetic fields, and is therefore called an electromagnetic wave. Unlike other waves with which people are normally familiar (such as sound or water waves), light does not require a medium to transmit through, but instead can travel through an absolute vacuum. Like other waves, electromagnetic waves have several measurable properties:

1. Light travels at a specific speed, called the speed of light. In a vacuum, this is about 186,000 miles per sec (300,000 km/sec). In his special theory of relativity (1905), Albert Einstein demonstrated that this is the ultimate speed limit for objects traveling in the universe (an inconvenience for cosmic travelers commonly ignored in science fiction).

2. Light waves come in different sizes, called wavelengths. The wavelength is the distance measured between two consecutive peaks (or

two consecutive valleys) in a light wave. The possible wavelengths of light vary from smaller than an atom to larger than a football field. The array of possible wavelengths is commonly called the *electromagnetic spectrum* as shown in Figure 1.1.

3. As a light wave travels by an observer, she can count the number of waves (cycles) that pass by her each second. This is called the frequency. If you multiply the wavelength of a light wave by its frequency, you always obtain the speed of light. Therefore, light with long wavelengths has a small frequency, and light of a short wavelength has a large frequency.

4. The height of the peaks and valleys of the wave gives the amplitude. The intensity of the light wave is related to its amplitude.

5. A final property of light is its *polarization*. Think about playing jump rope with a friend. As your friend holds one end still, you can shake the rope and set it into a wave-like motion. Depending on how you shake the rope, you can make the waves seem to travel side to side, up and down, or in any other direction (or plane) you wanted. If you carefully made the waves move in the same plane all the time, you will have polarized the rope waves in that direction. Some astronomical objects emit light polarized in a specific plane, while others emit light which is unpolarized (i.e., it can oscillate in any random direction).

Our first glimpse of the range of the electromagnetic spectrum began with Isaac Newton in 1671. When he passed seemingly white sunlight through a triangular chunk of glass called a prism, it was spread out into a rainbow of different colors which he projected on the wall. He named this effect the spectrum, after the Latin word for an apparition. What he really observed were various wavelengths of visible light, the portion of the electromagnetic spectrum which the human eye can directly detect. The different colors are bent by different amounts as they travel through the prism, and therefore spread out into their component wavelengths, or colors. Today both prisms and diffraction gratings (glass or other transparent material etched or colored with many close, parallel lines) are used to decompose light into its component wavelengths. The amazing amount of information astronomers can glean from such efforts (called *spectroscopy*) will be discussed in a following section. Visible light varies in wavelength from 400 (violet) to 700 (red) nanometers (nm), where there are 1 billion (10^9) nm in 1 m. For comparison, the average width of a single human hair is about 8,000 nm. It is no accident that our eyes are most sensitive to this part of the electromagnetic spectrum. The sun, our star, puts out the greatest amount of light at the yellow-green

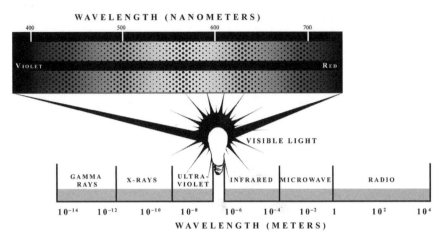

Figure 1.1 The Electromagnetic Spectrum

region of the spectrum (which is why it appears yellow to our eyes). This is also the portion of the spectrum where our eyes are most sensitive. Through the process of evolution, our eyes have developed to take advantage of the sun's most basic properties. Fortunately for astronomers, a great deal of the visible light from celestial objects passes through our atmosphere, allowing us to study our solar system, stars, and galaxies from the surface of the earth through this atmospheric "window."

In 1800, William Herschel found that a thermometer placed just beyond the red end of a visible spectrum would warm up, demonstrating the presence of energy at wavelengths just slightly too long for the human eye to see. He named these calorific rays, but today we refer to this as the infrared (IR) portion of the electromagnetic spectrum. Objects that are not hot enough to glow, but still radiate heat (such as human bodies and blacktop parking lots on a summer day) emit these wavelengths of light. They range from about 700 to 100,000 nm, or 0.7 to 1000 micrometers (μm). Night vision goggles and so-called heat-seeking missiles rely on detecting IR radiation. Heat lamps used to keep food warm at restaurants also utilize these wavelengths. Stars cooler than the sun, planets, and clouds of dust, especially those shrouding baby stars, naturally emit IR. Some wavelengths in this range reach the ground, but the longer ones are absorbed in the atmosphere, especially by water vapor, so IR astronomy is best done from mountaintops, balloons, or satellites above the atmosphere.

Slightly longer than IR waves are microwaves, which occupy the 1 mm through 3 cm portion of the electromagnetic spectrum. Although we normally connect microwaves with popcorn and nachos on demand, the universe itself emits important wavelengths in this range. This *cosmic microwave background,* accidentally discovered in 1965, is a ghostly relic from the early history of the universe, and provided important evidence for testing the original big bang model and its subsequent revisions.

The longest wavelengths of the electromagnetic spectrum are those referred to as radio. This encompasses a variety of human technologies, including AM, FM, and shortwave radio, radar, and television. The longest wavelengths considered useful by astronomers are in the range of 300–500 m (the length of several American football fields). The centers of certain galaxies emit a tremendous amount of radio waves, leading astronomers to hypothesize the cause to be black holes a million times the mass of our sun voraciously consuming entire stars and clouds of gas and dust. Although the existence of radio waves was predicted by James Clerk Maxwell in the mid-1800s, they were not produced in the laboratory until 1888 (by Heinrich Hertz), and celestial radio waves were not detected until 1931. With radio waves, we reach the end of the long wavelength, low frequency, or "red end" of the electromagnetic spectrum.

Returning to visible light, let us explore those wavelengths just a bit too short for the human eye to see. In 1801, J. Ritter discovered that silver chloride turned black when exposed to the rays of the sun's spectrum just beyond the violet wavelengths. This invisible radiation is now called ultraviolet (UV). With wavelengths between 1 and 400 nm, these rays interact with human tissue to cause sunburns, skin cancer, wrinkles, and cataracts. Fortunately for life on Earth, the stratospheric ozone layer absorbs a significant amount of the UV produced by the sun. However, destruction of the ozone layer through human activity (such as the creation of CFCs, or chlorofluorocarbons, the spraying of certain fertilizers, and jet contrails) is increasing the amount of UV light that reaches the ground, resulting in a complementary increase in skin cancer rates. In space, UV rays are produced by objects significantly hotter than the surface of the sun, such as massive stars, newly formed white dwarfs, and the centers of galaxies. Because our ozone layer absorbs UV, astronomers must rely on satellites to study these wavelengths.

In 1895, physicist Wilhelm Roentgen accidentally discovered invisible rays that could pass through most human soft tissues and allow the bones of living beings to be studied from outside their bodies. In 1901 he was awarded the Nobel Prize for discovering what he called x-rays. While the benefits of emergency room x-ray tests and the occasional dental x-ray are without dispute, repeated exposure does increase risks for cancer (hence the lead apron and questions about possible pregnancy that accompany x-ray photography, and why the technician leaves the room or stands behind a screen during the actual test). Fortunately for life on Earth, ozone (O_3) and molecular oxygen (O_2) in our atmosphere absorb most of the x-rays emitted by the sun and other celestial objects. By the same token, astronomers who wish to study x-rays emitted by the sun's million-degree corona, hot gas in clusters of galaxies, or gas being devoured by neutron stars or black holes, must rely on satellite observations. The wavelengths of x-rays are very small, ranging from 0.01 to 1 nm. In comparison, the diameter of an atom is about 0.1–0.5 nm.

The shortest waves in the electromagnetic spectrum are the gamma rays (γ rays), so named because they were the third type of radiation discovered to be emitted by nuclear reactions. Physicist Paul Villard accidentally discovered them in 1900 while experimenting with radioactivity. With wavelengths smaller than an individual atom ($10^{-7}$to 0.01 nm), these rays can be used to sterilize medical equipment and food (by killing bacteria) and can be used to kill cancer cells (in so-called "radiation treatment"). However, uncontrolled exposure can also kill healthy tissue, leading to radiation sickness, cancer, and death. Once again, Earth's atmosphere protects us from these harmful rays, but prevents astronomers from making ground-based observations. Satellites allow astronomers to study γ rays from solar flares and the centers of galaxies. At this point, we have reached the end of the short wavelength, high frequency, or "blue end" of the electromagnetic spectrum. Looking back on this survey of the electromagnetic spectrum, it becomes obvious that astronomers must observe the universe at all possible wavelengths in order to learn as much as possible about various celestial bodies. At the same time, astronomers must do most of these observations from space because of the shielding nature of our atmosphere in absorbing many wavelengths of light (especially the dangerous ones!)

A natural question to ask at this point is why are the harmful forms of light clustered in the short wavelength end of the spectrum? The answer was discovered by physicist Max Planck around 1900. He found that light has a dual nature—although we have been characterizing it as a wave, it can also act as a particle (which we call a photon). This wave-particle duality is one of the basic tenets of *quantum mechanics*, the branch of physics which studies the realm of the very small, including atoms. Planck found that there was a relationship between the wavelength of a light wave and the energy its photon carried. The shorter the wavelength (or higher the frequency), the bigger the punch the photon can deliver. Therefore a stream of photons of radio light can bombard your skin (and constantly does!) and you will not suffer any ill effect. However, a small number of γ ray photons carry enough energy to break apart DNA and can cause cells to become cancerous or die.

MAGNITUDES

Returning to visible wavelengths, it is obvious to even the most casual stargazer that stars differ from one another in brightness. This apparent brightness depends on two important factors—the star's distance and its energy output. The original scale used to classify stars by their apparent brightness was developed by the ancient Greek astronomer Hipparcos and popularized centuries later by Ptolemy in his naked-eye catalog of over 1,000 stars, published circa A.D. 140. The stars were divided into six classes, or magnitudes, with the first magnitude stars being the brightest. After the invention of the telescope, the magnitude scale was expanded to include seventh, eighth, and dimmer magnitude stars. In the eighteenth century, several difficulties with this antiquated scale became impossible for astronomers to continue to ignore. Unfortunately, as is commonly the case with astronomers, rather than throw out the old system and develop an improved and more logical classification from scratch, it was decided to patch up the old scale and make it workable, becoming the eternal bane of subsequent generations of astronomy students.

One of the difficulties was that the objects contained in the first magnitude represented far too great a spread in apparent brightness. Just as a homeowner might knock out a wall to expand a room, astronomers knocked out a mathematical wall, and expanded the magnitude scale to the left on a number line, first to magnitude zero and then to negative numbers. Therefore, the more negative the magnitude of an astronomical object, the brighter it appears. For example, the sun is magnitude −26.7, the full moon is −12.6, Venus is −4.4 at its brightest, and Sirius, the brightest star in the night sky, is −1.4. The second problem was that the difference in brightness between two consecutive magnitudes wasn't consistent. Since the average was about a factor of 2.5, it was suggested by astronomer Norman Pogson that the difference between magnitudes be standardized to 2.512, the fifth root of 100. Therefore, a magnitude 1.5 star appears 2.512 times brighter than a magnitude 2.5 star, while a magnitude 2.7 star appears $(2.512)^5$ or 100 times brighter than a magnitude 7.7 star.

So far we have only discussed how bright a star appears, or its *apparent magnitude*. As previously noted, this depends on both the distance and true energy output (luminosity) of the star. The intensity of light decreases as it moves farther from its source, according to an inverse-square law. For example, since Jupiter is about five times farther from the sun than Earth, the sun appears 5^2 or 25 times dimmer from Jupiter as compared to Earth. Since Pluto is about 30 times farther from the sun than Earth, the sun appears 30^2 or 900 times dimmer. So a star that appears dim in our sky might actually put out more energy than the sun but be located quite far away. In order to discuss the actual energy output of a star, astronomers introduce the concept of *absolute magnitude*. This is how bright a star would appear if it were located exactly 10 pc (32.6 light years) away. For example, although the sun's apparent magnitude is a whopping −26.7, its absolute magnitude is a disappointing +4.8, too dim to see from many light-polluted suburban skies.

Since absolute magnitude is defined with distance in mind, by comparing a star's apparent and absolute magnitudes we can determine its distance using a vitally important equation called the *distance modulus formula*. For readers intrigued by mathematics, the exact relation is $M - m = 5 - 5\log D$, where M is the absolute magnitude, m is the apparent magnitude, and D is the distance to the star, measured in parsecs. Therefore, if astronomers can somehow determine the true brightness of a celestial object, its distance can be calculated. This is a very powerful technique which will become the basis for measuring much of the universe.

TELESCOPES

As strange as it might sound, the facts surrounding the invention of the telescope are still under debate by historians. Credit is normally given to Johannes Lippershey, a Dutch optician, who applied for a license for the instrument on October 2, 1608. Soon after, Galileo Galilei heard about the invention and built one of his own, and became the first astronomer to make detailed telescopic observations. The name *telescope*, from the Greek *teleskopos* ("far-seeing"), was coined by Greek poet Ionnes Demisiani in 1612 (Webb 1999, 42).

Galileo's telescope was a refractor, whereas most professional modern telescopes are reflectors. Refraction is the bending of light as it passes from one material to another (such as air to glass, and air to water). This is why the part of a fishing pole submerged in a pond looks bent. This is also how a prism can take sunlight and spread it into its component colors. Different colors (wavelengths) are bent by differing amounts as they pass through a prism or lens. A telescope that uses a large (diameter, not thickness) lens to gather light is called a refractor. As seen in Figure 1.2, the light rays pass through the lens and are bent, cross at the focal point,

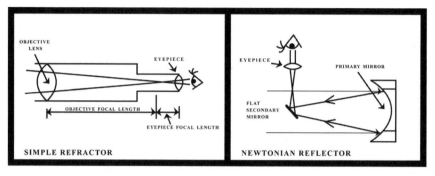

Figure 1.2 Basic Telescope Design

and continuing on to a second, smaller lens called the eyepiece. The light rays are bent once more and soon after pass into your eye, causing an image to be created on your retina. All that bending tends to make the image upside-down, which is not a concern to astronomers (since the universe has no real "up" or "down"), but can be quite disconcerting to the first-time observer when looking at the moon (or your neighborhood)! The distance between the center of a lens and the focal point is called the focal length, and is a fundamental property of that lens. For those readers who liked to burn leaves with a magnifying glass as a child, the focal length is the distance between the glass and the leaf, such that the sunlight appears most concentrated (and the leaf burn most easily). The magnification of the telescope (how many times bigger it makes objects appear) can be found by dividing the focal length of the large lens (called the objective lens) by the focal length of the eyepiece. To change the magnification, an eyepiece of a different focal length could be utilized. Theoretically, any telescope can magnify any number of times using the proper eyepiece (although useful eyepieces with focal lengths shorter than about one-half inch are increasingly harder to manufacture and tend to be costly). This must mean that there are other properties of a telescope which make it "better" than another.

Any telescope has three "powers"—light gathering ability, resolution, and magnification. The first two depend on the aperture, or diameter of the objective lens, with bigger definitely being better. As we have seen, magnification depends on the focal lengths of the two lenses. The most important of these three powers is the ability to gather light. A telescope is essentially a light bucket, gathering as much light as possible, in order to collect enough photons to make a distant object visible. If not enough light is gathered, no observation can take place! Just as a bigger bucket can gather more water, a bigger lens can gather more light, and creates a brighter image. Once an image bright enough to see has been made, the ability to see details, or resolution, becomes important. The smaller the detail that can be seen, the more information that can be gathered. Although in theory the resolution of a telescope depends solely on its aperture, in practice our atmosphere creates additional distortions which smear out the image of a star or galaxy and prevent fine details from being seen. Telescopes in satellites beyond our atmosphere avoid this problem. Once a bright, detailed image has been collected, its size can be manipulated through magnification. This is the easiest of the three powers to change, but is actually the least important. All too often, the first-time telescope buyer is only aware of magnification, and can be manipulated by salespersons accordingly.

Although well-made refractors give outstanding images, there are problems inherent in their design. As stated previously, different colors of light are bent by different amounts as they pass through a lens. Therefore, different colors will actually focus at slightly different points, causing the image of bright stars to take on a rainbow-like appearance. This is called chromatic aberration, and plagues cheap refractors. This can be overcome by using the proper combination of lenses, but it obviously increases the price of high quality instruments. The second problem is more serious. As anyone who wears eyeglasses can attest, the wider the lens, the heavier it is. But unlike eyeglasses, where plastics can be easily substituted, telescope lenses must be made of glass. This limits the size of the largest refracting telescope, for if a lens is too large, it will actually warp under its own weight. The largest refractor in the world is the 40-inch wide instrument installed at the Yerkes Observatory in Williams Bay, Wisconsin, in 1897.

In 1668 Isaac Newton invented a different type of telescope, one that utilized the reflection of a large mirror to gather light. The primary mirror of a reflecting telescope or reflector acts like the objective lens of a refractor, gathering light and determining the theoretical resolution of the telescope. Magnification, resolution, and light gathering ability are determined in a manner similar to a refractor. Reflectors have an additional complication. Because the light bounces off the primary mirror rather than passing through it, the focal point is located inside the tube of the telescope. This presents a problem, as stuffing one's head inside the telescope tube in an attempt to view the image prevents the image from forming in the first place! Instead, a small secondary mirror is added to send the light outside of the tube where it can be easily viewed. Different types of reflectors vary in how the secondary mirror is used. In

Newton's original design (shown in Figure 1.2), the light comes out of a small hole cut in the side of the tube, where the eyepiece holder and focuser are mounted. This Newtonian design remains the simplest to build (as this author can attest) and the most cost-effective telescope design, and is a favorite among amateur astronomers for these very reasons.

Because mirrors reflect all wavelengths the same, they do not suffer from chromatic aberration. In addition, because the weight of the mirror is more fully supported than in the case of a lens (which is held only at its edges), huge reflectors can be built. However, after a certain size, even with the support provided by the mount, the mirror will tend to warp. The largest unwarped primary mirror in the world is the 200-inch Hale telescope at Mt. Palomar Observatory outside San Diego (1948). Today, much larger telescopes can be made, either out of segmented mirrors (such as the 400-inch, or 10-m, Keck Telescopes in Hawaii) or with flexible mirrors equipped with special support mounts that reconfigure either the primary or secondary mirror's curvature as needed to compensate for the distortions. The same technology can be used to compensate for distortions caused by the atmosphere, a technique known as adaptive optics. The signals from several telescopes can also be combined in a process known as interferometry, resulting in an image which would otherwise have required a single telescope of unrealistic size to obtain. The largest telescope in the world, the VLT (Very Large Telescope) in Chile combines the signals from four 8.2-m reflectors, mimicking the images from a single 16-m mirror. It is important to note that if rigid mirrors are not carefully ground to the proper shape, they will not focus correctly. Such a spherical aberration plagued the Hubble Space Telescope until special corrective optics were incorporated into its cameras.

Modern astronomical research is no longer done by a human being directly looking through the eyepiece of a telescope (affectionately called "eyeball-on-glass"). Since the late 1800s, technology has enabled astronomers to record the light received from celestial bodies, allowing it to be studied at a later time and in greater detail. These technologies can also collect more light than the human eye, allowing for dimmer and dimmer objects to be studied. The first great technological advance of this type was obviously photography. The first photograph of a star was made in 1850 by J.A. Whipple at Harvard College Observatory. Over the next two centuries, photography greatly expanded our understanding of the solar system, stars, and galaxies. In a rather practical sense, it also opened the door to women in astronomy. As it was deemed improper and impractical for women to spend time in a dark, cold observatory, the fact that the photographs could be studied in the comfort and safety of a laboratory or library allowed women to work at major observatories for the first time in the late 1800s.

The charge-coupled device, or CCD, was invented in 1969 by Willard Boyle and George Smith at AT&T Bell Labs, again revolutionizing astronomy. These sensitive detectors use a computer chip to record almost every photon that strikes them, converting the information into an electrical signal which is processed by a computer. Using CCD technology, backyard amateur astronomers are now obtaining images of planets and galaxies superior to those obtained photographically a century ago by the world's largest telescopes.

Wavelengths other than visible light have their own types of telescopes, based on the principles of optics outlined above. For example, radio telescopes are generally constructed as large parabolic dishes that collect radio wavelengths and focus them to a detector. Radio astronomy can be done in the daytime as well as on cloudy days (at some wavelengths). Remember, though, that radio wavelengths are much, much longer than those of visible light, and it turns out that the resolution of a telescope depends on the specific range of wavelengths being collected. This means that radio telescopes must be huge in comparison to optical telescopes, and in practice don't achieve the same level of detail. Radio sources are also typically very faint, so the additional light gathering ability of a large telescope is required. The largest single dish radio telescope is the 1,000 ft Arecibo dish, built into a natural bowl-shaped valley in Puerto Rico in 1963, and was featured in the James Bond movie *Goldeneye*. The technique of combining signals from several telescopes (interferometry) was actually first used with radio telescopes.

The most important radio wavelength in astronomy is 21 cm, caused by cool clouds of hydrogen gas, or *HI regions*. An atom of hydrogen consists of one electron orbiting one proton. Each particle is said to possess the property of spin, which for the sake of simplicity we will (incorrectly) consider here as being like the rotation of the earth around its poles. The electron and proton can either spin in the same direction, or in opposite directions. When the two particles spin in the same direction (parallel), the atom has slightly more energy than in the case where they spin in opposite directions. The electron doesn't pick this higher energy state on its own, but can be coaxed into it by collisions with other particles. When the electron changes its direction (in order to return to a lower state of energy), it gives back the energy by emitting a photon of radio light with wavelength 21 cm. This special radiation is critical in mapping the structure of the Milky Way, where the presence of dust blocks visible light, making it otherwise impossible to see large parts of our own galaxy.

SPECTROSCOPY

Spectroscopy is the technique of intentionally breaking down the light from a source into its component wavelengths, using a prism or diffraction grating. It allows scientists to determine many physical properties of a light source, such as its composition, motion, and temperature. In 1814 Josef von Fraunhofer carefully analyzed the visible spectrum of the sun, and found that the rainbow of colors was interrupted by 10 prominent dark lines he dubbed A through H, and an additional 574 fainter lines between the B and H lines. This absorption spectrum became the basis of the chemical analysis of the composition of the sun.

In 1859, Gustav Kirchhoff developed three important rules governing spectra illustrated in Figure 1.3, all based on the observations of himself and Robert Bunsen:

1. A hot, glowing solid, liquid, or dense gas emits a continuous spectrum, consisting of a solid rainbow of all colors (wavelengths). An example is the metal filament in a standard incandescent light bulb.

2. A hot, thin gas emits light only at certain, particular wavelengths, creating an emission spectrum, and these wavelengths are unique to the composition of that particular gas.

3. If a continuous spectrum is passed through a cool, thin gas, the gas will absorb only certain wavelengths, creating an absorption spectrum. The wavelengths that are absorbed are identical to those emitted by the same gas when heated.

Because the wavelengths emitted and absorbed are unique to that chemical element, the presence (or absence, in the case of an absorption spectrum) of a particular spectral line in the light coming from a celestial object allows astronomers to determine its chemical composition. A set of spectral lines is as unique to an element as a set of fingerprints is to a person. In 1864, William Huggins was able to show that some nebulae— in his time a generic term for any object that appeared fuzzy when viewed in a telescope—produced emission spectra and are therefore hot clouds of gas in space. He was also able to show that they were comprised of the same elements that make up the sun. Spectroscopy yielded other surprises closer to home. For example, helium (named for Helios, the Greek god of the sun) was discovered in the spectrum of the sun in 1868, nearly 30 years before it was found on Earth.

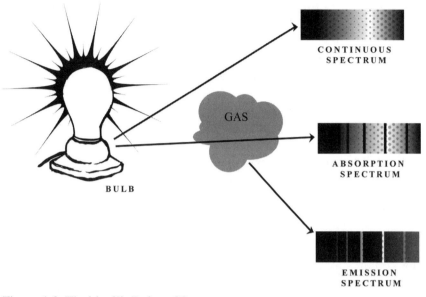

Figure 1.3 Kirchhoff's Rules of Spectra

In order to understand why spectroscopy allows for chemical identification, we must return to the field of quantum mechanics. In 1912, Danish physicist Niels Bohr developed a model of the atom that resembled the solar system. In this model, the negatively charged electrons orbit the positively charged nucleus in specific, allowed orbits, each with its own inherent energy. The simplest atom to consider is hydrogen. It also makes up about 75 percent of the universe (by mass), so it is a very important atom to understand. A single electron orbits a single proton due to the electrical attraction of their opposite (but equal) charges. Because of this attraction, the electron normally occupies the closest allowed orbit to the proton, known as the *ground state.*

In order to occupy other orbits which are farther from the nucleus, also called excited states, the electron must absorb energy (in the form of a photon). But not any old amount of energy will do! It takes a specific amount of energy for the electron to "jump" to the first excited state, and unless the electron absorbs a photon of just the right energy (with just the right wavelength), there will be no transition. The absorbed photon will create an absorption line in the spectrum of the light passing through this collection of atoms. Likewise, there is another specific amount of energy (and hence another corresponding wavelength of light) associated with a transition to the second excited state, the third, and so on. Since there are many, many possible excited states, there are many possible transitions, each with its unique associated wavelength of light. The spacing of the orbits (also known as energy levels) is fixed by the atomic properties of hydrogen, and is different for other elements, thus resulting in unique spectral lines being associated with every element.

Once in the excited state, the electron prefers to release its acquired energy and return to the ground state. In order to do this, it must emit a photon of light of the exact same energy (and wavelength) that it absorbed to make the jump in the first place. This creates the unique emission lines created by a cloud of hot, thin gas. Many electrons are excited due to the high temperature, and in order to return to the ground state they must emit energy, thereby creating the emission spectrum.

Some transitions are more easily accomplished under conditions found on Earth and are therefore more commonly seen. However, there are certain transitions that are very difficult on Earth, but much easier in the near-vacuum of space. These transitions were unexpected by astronomers and when discovered in the spectra of celestial objects were named forbidden lines. An important example is a green spectral line in the spectrum of oxygen atoms that have lost two of their eight electrons (doubly ionized oxygen). It was originally attributed to an unknown element dubbed nebulium, since it was discovered in the spectra of hot clouds of gas in space.

Among the contributions to quantum mechanics of physicist Max Planck was the realization that a truly continuous spectrum over all wavelengths is only possible for an idealized object known as a *blackbody*. This perfect emitter is also a perfect absorber, and hence should be utterly black when cold. Although blackbodies and their radiation are idealizations, stars are decent approximations (at least near the visible portion of the spectrum), as is the cosmic microwave background.

Blackbodies have three important properties that are of use to astronomers:

1. They emit light over a wide range of wavelengths, with varying amounts of energy at different wavelengths. The graphical representation of this is called the *Planck curve*.

2. They emit most intensely at one particular wavelength, which is related to their temperature. This results in a peak in the Planck curve. The hotter the object, the shorter the wavelength of maximum emission. This explains why hot stars are bluish, cooler stars are reddish, and middle of the road stars like our sun appear

yellow. The relationship is called *Wien's law*, and gives astronomers a way to estimate the surface temperature of the star.

3. The hotter and larger the blackbody, the greater its total energy output, or luminosity. The luminosity is related to its surface area (radius squared in the case of a sphere) and its temperature raised to the fourth power. This *Stefan–Boltzmann law* relates true brightness, surface temperature, and size of a star. If the first two properties are known, the third can be estimated.

A final property of stars and galaxies that can be gleaned from a study of light is motion toward or away from Earth (known as the *radial velocity*). In 1842, Austrian scientist Christian Doppler announced that the wavelength of any wave (be it a water wave, sound wave, or light wave) becomes shorter if the wave source were approaching the observer, and would become longer if the source were receding. This *Doppler effect* is most obvious for sound waves, where the change in wavelength creates a change in the pitch (not volume) of the sound. If an ambulance is approaching you, its siren becomes more shrill (its pitch increases) as its wavelength shortens, but when the same ambulance rushes away from you, the pitch lowers (becomes more bass) as the wavelength stretches out. Likewise, when a star or galaxy approaches Earth, its light is shifted toward shorter values, or toward the blue end of the electromagnetic spectrum. This is called a blueshift. A receding star or galaxy has its light waves lengthened, shifting toward the red end of the spectrum (termed a *redshift*). The amount of the shift directly depends on the speed of the object. Therefore, it is possible to calculate both the direction of the motion (toward or away) as well as the speed. For example, the bright star Sirius was found to be approaching Earth by William and Margaret Huggins in 1868 based on a 0.015 percent blueshift in its absorption spectrum, as compared to the sun's spectral lines (Singh 2004, 244).

Although astronomers passively receive the shifted light from objects in space, on Earth we can actively bounce wavelengths (usually in the microwave or radio part of the spectrum) off a moving body and by noting the change in wavelength of the light that returns, estimate the speed of the moving body. This is the basic principle behind the Doppler radar used by meteorologists to study the motion of precipitation inside clouds, as well as "radar guns" used by police to catch speeders. It is indeed amazing to think that the same technology that studies the spiraling formation of tornados can be used to examine the rotations of galaxies.

THE CLASSIFICATION OF STARS

Although many people are under the impression that all stars are white, a careful survey of the night sky will convince them that this is really not the case. Some stars appear yellowish, like our sun, while others exhibit varying intensities of blue, orange, or red. As we have already seen, the color of stars is determined by their surface temperature, through Wien's law. Stars also differ in size and true brightness (luminosity or absolute magnitude), with these properties related to the temperature through the Stefan–Boltzmann law. Astronomers use these properties to classify stars into various groupings, and then utilize this information to study their lives and deaths.

It bears repeating that most of what astronomers learn about stars is achieved by studying their spectra. This includes composition and motion. In this vein, Annie Jump Cannon lauded the universe's "Patient light! Shining down on humanity these countless centuries until man became clever enough to wrest from its vibrations the secrets so closely concealed" (1941, 56). It is therefore natural to ask if stars exhibit different types of spectra and if so, what inherent properties are responsible for these differences. The answer was first noted by Fraunhofer himself around 1814. Not only did he study the spectral lines of the sun, but he found that some bright stars had similar spectra, while others had spectra with differing appearances. The first significant classification system was done by Jesuit priest and astronomer Father Angelo Secchi in the 1860s. His equipment was crude by modern standards, and he was limited to making drawings at the eyepiece (as Fraunhofer had done), but he was still able to divide stellar spectra into five basic classes. Type I stars, including Sirius, were bluish-white in color (and therefore hotter than the sun). Their spectra exhibited stronger hydrogen spectral lines than did the sun (although Secchi did not understand this). Type II stars were similar to the sun—yellow stars of medium temperature whose spectra exhibited relatively strong lines now known to be due to calcium and other metals. Type III stars were cooler orange and red stars. Their spectra had curious bands—close forests of lines—near the blue end of the spectra. These are now known to be due to molecules of titanium oxide (TiO). Type IV stars were very red and had bands of lines due to carbon (which Secchi himself realized). Type V stars were very peculiar and rare, exhibiting strong emission rather than absorption lines.

Secchi's classification was very influential in his time, and further refinements required better technology—namely photographs of spectra which could be studied and repeatedly compared. The first photograph of a spectrum (of the bright star Vega) was taken by New York physician and amateur astronomer Dr. Henry Draper in 1872. After achieving this task, he and his wife, Anna, set a more ambitious goal of photographing and classifying stellar spectra in more detail than Secchi had been able to achieve. Unfortunately all those chilly nights in his personal observatory led to Draper's death in 1882 from pneumonia at age 45. After her husband's death, Anna Draper tried to carry on the project with a series of unsatisfactory assistants, but finally gave up, and after exchanging a series of letters with the Director of the Harvard College Observatory (HCO), E.C. Pickering, she decided to endow the HCO with sufficient funds to carry out the work in her husband's honor. The result was the Henry Draper Memorial, which produced the most detailed catalogs of stellar spectra ever compiled.

Pickering had begun hiring women "computers" in 1879, low-paid workers who did much of the labor-intensive calculations and measurements of the mountain of photographic plates generated by the various ongoing research projects at the HCO. With the funding provided by Mrs. Draper, Pickering was able to hire so many women that they became known as "Pickering's Harem." Beginning in 1886, the women computers at Harvard began measuring the positions of spectral lines for thousands of stars, using the sun's spectrum as a standard. The initial work was mainly done by Nettie Farrar, who left HCO at the end of the year to be married. She was replaced by Pickering's former maid, Williamina Paton Fleming, who, according to an often-repeated story, was originally hired as an observatory assistant by Pickering some years before in a fit of exasperation, complaining of his male assistants that his housekeeper could do a better job. Fleming became the supervisor of a number of women computers, and in 1898 became the first woman to receive an official HCO appointment, Curator of Astronomical Photographs. The first Henry Draper Catalog was published in 1890 and contained the classified spectra of 10,351 stars. Pickering and Fleming had developed a new system of classification in the preparation of this work, an alphabetical system from A to M (deleting J, to avoid confusion with I) which listed the spectra in order of increasing complexity, and decreasing relative strength of the hydrogen lines. Letters N–Q were reserved for stars of peculiar spectra.

By this time, Pickering had already begun taking advantage of technological improvements and organized more detailed evaluations of stellar spectra. In 1888 he hired Draper's niece, Antonia Maury, to reclassify stars in the northern half of the sky utilizing this improved technology. Maury, a graduate of Vassar's astronomy program (under the direction of Maria Mitchell, the first American woman astronomer), bristled under Pickering's orders to merely improve Fleming's system, and instead developed her own unique classification. Not only did she separate the spectra into 22 numerical classes, but she reordered some of Fleming's classes, putting O and B before class A. More importantly, she rated the spectra on a second dimension, the appearance of the spectral lines themselves. Normal spectral lines were class a, while class b spectral lines were hazy, and class c were unusually narrow. Maury's catalog and system were eventually published in 1897, but went virtually ignored for a decade. The HCO instead officially sanctioned the system devised by fellow Harvard computer Annie Jump Cannon.

Cannon, a Wellesley physics graduate, was hired by Pickering in 1896 to classify the southern stars. She agreed with Maury's rearrangement of the original Fleming classes, but ignored Maury's line-width scheme. She also omitted classes which had been shown to be artifacts of the photographic plates, resulting in the now famous spectral system OBAFGKM, often remembered by astronomy students through the rather politically incorrect mnemonic device "Oh Be A Fine Girl Kiss Me." The system was later refined by Cannon to include subtypes denoted by the numbers 0–9 within each spectral class. It should be noted that over the course of her four decades at the HCO, Cannon personally classified the spectra of over 350,000 stars. Her classification system appeared in 1901, and was endorsed by the International Astronomical Union in 1922. It was ordered that any future extensions or changes utilize her system as their foundation. In the past decade, the spectral classification scheme has been accordingly expanded to include even cooler objects, the L (1997) and T (1999) classes. Approximately half the L class objects are true stars. The rest, and all the T objects, are *brown dwarfs*, failed stars that began life with too little mass to generate energy via nuclear fusion.

Cannon's classes list stars in order of decreasing surface temperature (as determined by their color), although the actual temperatures corresponding to each spectral class were determined by Cecilia Payne-Gaposchkin in 1925 as part of her ground-breaking Ph.D. thesis. It was the first thesis completed in astronomy at Harvard, and in it she also demonstrated that the overall composition of most stars is similar—approximately 75 percent hydrogen and 25 percent helium (by mass). The idea that the relative strength of spectral lines of different elements in a star is due to temperature and not composition may seem contradictory, but actually follows directly from the same concept of energy levels in an atom that gives rises to the lines in the first place. Since the spacing of the energy levels varies from element to element, there is some range of temperatures at which the electrons of a given element most efficiently and most often jump to excited energy states. However, if the energy (temperature) is too great, electrons can actually be removed from the atom, called ionization. This will weaken the relative strength of the excited state spectral lines. If the energy (temperature) is too low, the electrons won't make many jumps at all, and the strength of the spectral line will also be weaker.

THE HR DIAGRAM

But what of Maury's second dimension? In 1905, Danish astronomer Ejnar Herzsprung discovered that red stars come in two very different varieties—low luminosity dwarfs located in our astronomical neighborhood, and high luminosity giant stars which can be seen over large distances. Interestingly, these red giants were exactly the same stars which Maury had labeled with her "c-characteristic." Herzsprung wrote to Pickering, calling her discovery,

the most important advancement in stellar classification since the trials by Vogel and Secchi. . . . To neglect the c-properties in classifying stellar spectra, I think, is nearly the same thing as if the zoologist, who has detected the deciding differences between a whale and a fish, would continue classifying them together. (Jones and Boyd 1971, 240)

Pickering countered that the spectra did not show enough detail to verify Maury's claim. During the same period of time, the preeminent American astronomer Henry Norris Russell had independently discovered that stars could be found in different size types. It was Russell who first plotted a diagram demonstrating this, and today the *Hertzsprung–Russell diagram* forms the backbone of stellar astronomy. What Maury had unknowingly discovered was that swollen giant stars have lower atmospheric pressures than smaller stars, resulting in the different appearance of their spectral lines.

The HR diagram (shown in Figure 1.4) is generally plotted using spectral class, temperature, or some measure of a star's color on the horizontal (*x*-axis), and luminosity, absolute magnitude, or some other measure of a star's intrinsic brightness on the vertical (*y*-axis). Dim stars are plotted near the bottom, bright stars near the top. If absolute magnitude is used, this means that large positive numbers (such as +15) are near the bottom and large negative numbers (such as −7) are near the top. Hotter stars appear on the left side of the diagram and cooler stars toward the

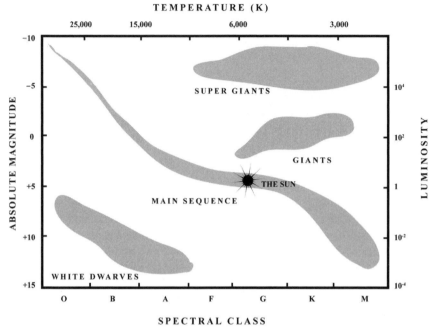

Figure 1.4 The Hertzsprung-Russell (HR) Diagram

right. It is found that over 90 percent of stars appear on a fairly narrow swath starting in the upper left corner (bright and hot) and moving toward the lower right corner (dim and cool). This is where normal stars are found, those that Hertzsprung called dwarf stars, more commonly known as *main sequence stars*. Note that these stars are not necessarily small, as our sun belongs to this category; they are, however, much smaller than the giants. If our sun were replaced by a red giant, it would swallow up the orbits of all the planets through Earth. Stars not found on the main sequence are dying stars. For example, a small clump of stars are located in the lower left section of the HR diagram. These are very hot, yet very dim, meaning they must be small in size. These are stellar corpses approximately the same mass as the sun, called *white dwarfs*. Stars in the upper right part of the diagram are relatively cool, but very bright, meaning they are large in size. This is where the giants and supergiants dwell.

A detailed study of the HR diagram and the various size classes of stars was published by William Morgan and Philip Keenan, assisted by Edith Kellman, at Yerkes Observatory in 1943. The resulting MK or Yerkes system was a two-dimensional system, like Maury's, which introduced a luminosity class in addition to the regular spectral class. The luminosity classes are designated by Roman numerals, and are as follows:

Ia	bright supergiants
Ib	supergiants
II	bright giants
III	giants
IV	subgiants
V	dwarfs (main sequence)
VI	subdwarfs

For example, the sun is a G2V star. The reason for the large number of luminosity classes was first hinted at in the work of Nancy Roman at Yerkes Observatory. She noticed that some stars had very few elements heavier than hydrogen and helium, although their surface temperatures appeared the same as "normal" stars. This was some of the first evidence that stars occur in different populations by composition.

THE INTERSTELLAR MEDIUM

Interstellar space is commonly thought of as being completely devoid of matter. Fortunately this is not the case, for without the interstellar medium (ISM) new generations of stars would not be possible. Clouds of gas and dust are generically termed *nebulae*, and are commonly associated with either the birth or deaths of stars. They form a sort of astronomical Rorschach inkblot test, and often carry whimsical names such as the Horsehead, the Crab, and the Cat's Eye.

Interstellar dust grains are believed to be composed of carbon (graphite) and silicates, the former being created in the atmospheres of a particular type of cool supergiants formerly called carbon stars (Secchi's Type IV stars). Because the size of these dust grains is in the same range as the wavelengths of visible light, light interacts with the dust in interesting ways. Light from distant stars is generally dimmed by the intervening dust, an important effect called *extinction*. It causes stars and other celestial objects to appear dimmer than they should, introducing error into distance calculations which rely on a comparison of apparent and absolute magnitude. If a star appears dimmer than it should, it will be assumed to be more distant than it truly is. Blue light is preferentially scattered from the dust because of the size of its wavelength, resulting in two complementary effects. First, blue light from hot stars scatters off clouds of dust, turning these reflection nebulae blue. Second, since the light from the star has lost some of its blue component, the starlight is "deblued," or appears redder than it should. Like extinction, this *interstellar reddening* must be taken into account when studying the properties of stars. If the dust is dense enough, it will prevent any visible light from traveling through, creating a dark nebula which is usually only seen when positioned in front of a luminous nebula.

The other component of the ISM is gas, mainly hydrogen, either neutral (cool) or ionized (hot). *HI regions* are clouds of cool hydrogen which are detectable via the 21 cm radiation they emit. *HII regions* are reddish clouds of hot hydrogen. Ultraviolet light from nearby hot, young stars ionizes the hydrogen. The electron is eventually recaptured by its nucleus (proton), gradually winding its way back to the ground state. As the electron falls from the third energy level to the second it emits a characteristic red wavelength called hydrogen alpha, which is responsible for the red color of these emission nebulae. *Planetary nebulae* are shells of gas emitted from dying stars, while supernova remnants are the shredded remains of exploded stars, returning enriched material to the ISM for recycling into a new generation of stars.

Approximately half of the mass of the ISM is contained in dense clouds composed mainly of molecular hydrogen (H_2). These molecular clouds also contain over 100 different compounds including water, formaldehyde, alcohol (both the rubbing and drinking varieties), sugar, and even amino acids, the building blocks of life. Exceedingly cold (only 5 degrees above absolute zero), these gigantic blobs of gas can contain 10 million times the mass of the sun in a diameter of dozens of parsecs. Molecular clouds are the stellar nurseries in which new stars form. But how can a diffuse, distended cloud of gas become a compact nuclear inferno?

THE LIVES OF STARS

Star Formation

Star formation is believed to be kick-started by some sort of trigger, perhaps the shock wave from the supernova explosion of a nearby dying star, or the intense stellar wind rushing out from neighboring hot and bright O- and B-type stars. A small knot of material within the cloud is compressed, and gravity takes over. As the forming *protostar* collapses its atoms collide, generating greater and greater heat. In order to conserve angular momentum, any initial rotation the protostar has also increases, just as a figure skater spins faster by pulling in her arms. This creates a serious problem for the would-be star. Unless it finds a way to shed a significant portion of its angular momentum, calculations show that a newborn star would spin faster than the speed of light, which clearly is not allowed. By comparison, as Galileo discovered, the sun spins roughly once a month. The key is a protostar's magnetic field, which can act like a brake, slowing down its rotation. Otherwise, the protostar might be ripped apart by its own spin! As the protostar continues to contract and heat up, a disk of material forms around it, which can eventually form planets, as in our own solar system. In fact, there is so much excess material associated with star formation that it typically takes place shrouded in a cocoon of dust and gas, hiding it from direct view until the star's stellar wind turns on and blows it away. Astronomers normally study protostars through the IR produced from the cocoon, which itself is warmed by the protostar it contains.

When the protostar's core temperature reaches about a million Kelvin (10^6 K), fusion occurs for the first time. Fusion is a nuclear reaction in which small atomic nuclei are combined to form a larger one, releasing a tremendous amount of energy. Today's thermonuclear weapons (hydrogen bombs) are an uncontrolled example. *Back to the Future's* Doc Brown may have safely harnessed fusion to power his time machine, but real scientists have not been so fortunate. The original atom bombs and nuclear power plants are examples of a different type of nuclear reactions called fission. In this process, a large nucleus, such as uranium or plutonium, is split into smaller pieces, also releasing energy.

In the protostar's first bout of fusion, deuterium, or heavy hydrogen, fuses to form helium-3, a light form of helium with a nucleus made of two protons and one neutron. This temporarily halts gravity's collapse of the protostar, and in fact the object temporarily swells to several times its final size. But deuterium is quite rare, so this phase in the protostar's life is brief by stellar standards. Once the deuterium in the core is exhausted, gravity returns to center stage, and the protostar resumes its collapse. If it is about sixty times the mass of Jupiter, or 6 percent the mass of the sun, its core reaches the 2×10^6 K necessary to fuse lithium, the third element on the periodic table. Once again, the collapse is temporarily halted for a few million years, until the protostar's limited supply of this rare material is exhausted as well.

The collapse now resumes under gravity's relentless urging, and if the protostar is smaller than 75 times the mass of Jupiter (7.5 percent the mass of the sun), no further fusion cycles are possible. Its core simply cannot generate enough heat to overcome the natural repulsion of positively charged protons for each other. Such objects are failed stars, or *brown dwarfs*. They continue to collapse until the electrons get as close together as they possibly can, filling up all the lowest energy states. This *degeneracy pressure* holds up the brown dwarf against further collapse. Once formed, the brown dwarf can do nothing more than become cooler and dimmer over time.

A happier fate awaits protostars above the critical mass limit of 0.075 solar masses. When the core finally reaches about 10^7 K, protons (hydrogen nuclei) are moving fast enough to overcome their normal electrical repulsion for each other, and instead fuse together to form helium, releasing considerable energy in the process. The energy stabilizes the interior against the pull of gravity, and a star is born. It is at this point that the new star occupies a point on the main sequence of the HR diagram. The zero age main sequence (ZAMS) is a line on the HR diagram representing the original position of stars of different masses, with the most massive at the top left and the least massive on the bottom right. The more massive the star, the higher its temperature, and the faster it uses up its fuel to generate that energy, resulting in a greater brightness. The downside is that it dies sooner.

Our discussion has focused on single stars, but it is clear that any molecular cloud contains enough material to create many, many stars. About half of all sun-like stars and one quarter of lower mass stars are members of binary systems, in which two stars orbit each other. Multiple births in the form of triplets and quadruplets of stars are, as in the case of humans, much less common. Larger groupings of stars, with dozens or even thousands of members, also exist. These are OB associations, open clusters, and globular clusters. *OB associations* are, as their name implies, loose groupings of dozens of very hot and short-lived O and B stars, spread across 100 pc. The most famous example is the stars of Orion's belt. The mutual gravitational pull of the individual stars is not sufficient to hold the association together, and they tend to fall apart in only a few million years (assuming that their member stars don't die first!). *Open clusters* are also irregularly shaped, but include from several hundred to a thousand fairly young stars within a diameter of several dozen pc or less. The Pleiades, or Seven Sisters, in Taurus is a classic example. Open clusters are more tightly bound than OB associations, but stars can still wander away over time as the cluster is tugged upon by other stars or giant molecular clouds. OB associations and open clusters are sometimes called galactic clusters, as they are found in the spiral arms of the Milky Way.

Globular clusters represent a very different sort of cosmic critter. These are spherical swarms of hundreds of thousands to a million older stars packed into a diameter of less than 50 pc. In a typical backyard telescope they have the general appearance of a small cotton ball, with some individual stars resolvable at the edges. Most globular clusters occupy a more or less spherical halo around the plane of our galaxy, and represent some of the oldest stars known. Because of their age, globular clusters are important laboratories for the study of stellar evolution.

Stellar Evolution

When gazing skyward, it is easy to believe, as the ancients did, that the stars are eternal and unchanging. But as was sometimes noted, stars can dramatically change in brightness, and in doing so seemingly appear and disappear from view. The ancient Chinese noted the occasional appearance of "guest stars," while Renaissance astronomer Tycho Brahe coined the term "nova," the Latin term for "new." Despite these rare yet spectacular events, the evolution of an individual star is generally so slow that a single generation of astronomers cannot hope to see much in the way of change in the properties of any individual star. How, then, can astronomers claim to understand the evolution of stars from cradle to grave with any accuracy? The key is the large number of stars we have to observe, each at a slightly different point in its life cycle. Consider the case of a group of ant botanists. Compared to the lifespan of an individual tree, a single ant lives for a trivially short period of time. But by observing many trees of different species at different points in their development—seeds, seedlings, saplings, mature trees, dying trees, and fallen dead trees—the ant scientists could piece together how trees are born and die. Astronomers follow the same procedure in studying stars.

Main sequence stars represent the stable portion of a star's life. It is in hydrostatic equilibrium, where the inward pull of gravity is perfectly balanced by the outward push of the energy it generates via nuclear fusion. A main sequence star converts hydrogen into helium in its core. Because stars are 75 percent hydrogen and 25 percent helium (by mass), it is tempting to think that by the end of a star's life it becomes 100 percent helium. But note the previous phrase "in its core." Only in the core of the star is the temperature high enough for fusion to occur. This means that unless the star can effectively circulate its material (called convection), only the hydrogen in the core is available for fusion. Very low mass stars (M and L class main sequence stars under 0.4 solar masses) are completely convective, able to circulate their material from core to surface. Not only are they exceedingly fuel-efficient in that they are miserly in their energy generation and resulting brightness, they are also able to use all the fuel at their disposal. These underachievers of the stellar world may not dominate our visible night sky, but they represent the vast majority of all stars in our galaxy, and their estimated lifespan is 10 trillion years—nearly 1000 times the current age of the universe. As you might expect, we will revisit this startling concept. Main sequence stars over 0.4 solar masses are limited to using only the hydrogen initially in their cores, as convection only occurs in the outer portion of the star near the surface.

To quote an Iron Maiden song, "as soon as you're born, you're dying." For stars this is most certainly true. From the moment of its first appearance on the ZAMS, the very act of shining changes the chemistry inside a star. There are two general fusion cycles which stars can use to convert hydrogen nuclei into helium nuclei. Stars with masses similar to the sun and lower have cooler core temperatures, relatively speaking, and rely on the *proton–proton cycle* to generate energy. Although there are several branches of this reaction, each with several steps, the main idea is that four hydrogen nuclei (four protons) fuse to form one nucleus of helium (also called an alpha particle), which is composed of two protons and two neutrons. In addition, several other particles, including neutrinos, are produced. This process generates a tremendous amount of energy through the most famous equation of all time: $E = mc^2$. Recall that the speed of light (c) is a rather large number—300,000 km/s. The equal sign in the formula means that energy can be converted into mass, and vice versa, with the speed of light squared as the conversion factor. Because of this, a small amount of mass can be converted into a tremendous amount of energy. This is similar to certain currency exchanges; for example a few dollars can be converted into an impressive number of rupees. The helium produced by the fusion (and the very light waste products) has less total mass than the four protons that went into the reaction, with the excess mass being converted into energy. As a result, the sun and all main sequence stars are constantly losing mass as they generate energy. For example, the sun is losing over 4 million tons of mass every second! At its present rate of energy production, the sun is not quite halfway through its estimated 10- to 11-billion-year main sequence lifespan.

With a core temperature of about 15 million Kelvin, the sun generates 90 percent of its energy using the proton–proton cycle. The other 10 percent is due to the *carbon or CNO cycle*. Stars with several times the sun's mass or greater are able to exploit more fully this more efficient fusion cycle, which utilizes carbon as a catalyst or agent of change. The CNO refers to an overview of the cycle, in which a carbon nucleus is converted into nitrogen, oxygen, and then back to carbon again. In the process, four protons are again converted into one helium nucleus, more energy, and light waste particles. In order for the carbon cycle to be effectively utilized by a star, the core must be hotter (preferably 20 million Kelvin or more) in order for the reactions to take place, meaning a higher mass star, and of course carbon must initially be present.

For millions or billions of years (depending on its mass), a star happily churns out energy through the fusion of hydrogen into helium. But when all the hydrogen in the core is exhausted, the main sequence portion of a star's life is spent. The details of what happens next depends, not surprisingly, on the mass of the star. First we will consider a low mass star, defined as eight solar masses or less. When hydrogen fusion ceases in its core, there is nothing to keep gravity at bay, and the helium core begins to contract and heat up. This extra heat raises the temperature of a shell of hydrogen surrounding the core to the requisite 10 million Kelvin and hydrogen begins fusing into helium. But this is no cause for celebration. The star is forever banished from the main sequence, and this is just part of the star's dying process. The interior of the star is not stable, and the outer layers expand out into space, swelling the star's size to as much as the diameter of Earth's orbit. As the star's outer layers move farther from the hydrogen fusion they cool, and take on a reddish tinge. This is the red giant phase of the star's life, and it occupies a position on the HR diagram called the *red giant branch* (as shown in Figure 4.7 on p. 104).

As the core continues to collapse and heat up, it eventually reaches another critical temperature. At 100 million Kelvin, helium fuses to form carbon. In this *triple-alpha cycle*, three helium nuclei are converted into one carbon nucleus and energy. An additional helium nucleus can also combine with the carbon to form a nucleus of oxygen. Hydrostatic equilibrium is reestablished inside the star, and it retreats from the red giant portion of the HR diagram, taking a location on the *horizontal branch*. There it remains, for about 20 percent of the amount of time it had spent on the main sequence, until the helium in the core is exhausted. Gravity regains the reins, and for the second time the star becomes a red giant. The increased core temperature converts the shell of helium created in the hydrogen shell burning stage into a new source of fusion (helium shell burning), while a shell of hydrogen just outside this is now hot enough to begin a second episode of hydrogen shell burning. During this double shell burning stage the star occupies the *asymptotic giant branch* of the HR diagram. The star's mass is too low for its gravity to compress it enough for temperatures necessary to fuse carbon and oxygen to be reached, so this is the end of the road. The tenuous outer layers of the star are repeatedly puffed off into space, forming a *planetary nebula*, and the core contracts until the degeneracy pressure of its electrons halt the collapse, forming a *white dwarf*. It should be explained that the term planetary nebula is yet another example of astronomical obfuscation. The name does not imply that the objects are related to planets, but rather refers to their roundish shape, which reminded nineteenth century astronomers of planets.

Recall that low mass red dwarf stars (under 0.4 solar masses) are fully convective and can convert all their hydrogen into helium. But at the same token they never achieve high enough temperatures to convert helium into carbon, nor do they undergo shell burning. Therefore they never become red giants, but instead slowly become brighter and hotter as they use up their hydrogen stockpiles. Near the end of their main sequence lives they may actually look very similar to our sun, but only for a brief moment, in cosmic terms, before collapsing to form a white dwarf.

We now turn our attention to massive stars, more than eight times the mass of the sun. Unlike their lighter cousins, these heavyweights generate high enough temperatures in their cores to fuse carbon and oxygen into heavier elements. In fact, the heavier the star, the higher its possible core temperatures, and the heavier the elements it can fuse. These stars will reach the red supergiant section of the HR diagram in their later evolution, swelling to a thousand times the diameter of the sun. If our sun were replaced by one of these behemoths, it would easily engulf the orbits of Mars, the asteroid belt, and in some cases even Jupiter. Once helium fusion completes, the star permanently becomes a red supergiant, as each new fusion cycle lasts for a briefer period of time than the last, and the exterior of the star does not have sufficient time to readjust. Elements through iron are created as the star dances its way down the periodic table, creating the elements necessary for the existence of planets and life. But iron is the end of the road as far as fusion cycles go. It takes more energy to coax iron to fuse than is liberated in the fusion reaction. So how were the heavier elements in the universe created?

The problem is that heavy elements have a large number of protons, and their large collective positive charge strongly resists any other protons (or helium nuclei) getting too close. Neutrons, being neutral, are not rejected. Nature takes advantage of this through the *s-process* and *r-process*, named slow and rapid for their relative speeds. A heavy nucleus can capture as many neutrons as it wants, until it becomes radioactively unstable, and decays into a different element, perhaps one that cannot be created inside stars by any other means. The s-process occurs in the outer layers of red giant stars, and is a back-door method for low mass stars to contribute to the chemical evolution of the universe. Once the new heavy elements are created, they are blown into space in the planetary nebula, enriching the ISM and the next generation of stars. The r-process occurs in the explosion of stars, called a *supernova*. The study of how chemical elements are created inside stars is called *stellar nucleosynthesis* and was pioneered in a seminal paper published in 1957 by E. Margaret and Geoffrey Burbidge, William Fowler, and Fred Hoyle. As we shall see, in the early history of the universe only hydrogen and helium were created. Every atom of oxygen you breathe, every atom of gold you wear, and every atom of carbon in your cells was created inside a dying star that existed long before the sun. In the famous words of astronomer Carl Sagan, "We are star stuff."

Cepheids and the Period-Luminosity Relationship

During a star's death throes, when it is moving between fusion cycles, it crosses an area of the HR diagram called the instability strip. As the name implies, stars in this section of the HR diagram are inherently unstable, and will actually pulsate. Two important types of pulsating variable stars can be found here, *RR Lyrae stars* and *Cepheids*. RR Lyrae stars are lower mass stars that pulsate in a day or less. Cepheids, named after Delta Cephei, the first example discovered, are higher mass stars located further up on the HR diagram, and take several to dozens of days to complete one cycle of variation. Stars that vary in brightness for any reason are monitored, and once sufficient data is gathered, the apparent magnitude versus time is graphed, leading to a pattern called a light curve. The shape of the light curve is unique to that type of variable, and is used by astronomers to understand why and how that particular type of star varies, by comparing it to computer models.

The women computers at the Harvard College Observatory not only classified stellar spectra a century ago, but also discovered and monitored thousands of variable stars using photographic plates. Among the most prolific of the variable star discoverers at Harvard was Henrietta Swan Leavitt. One of her studies involved identifying variable stars in the Large and Small Magellanic Clouds, now known to be small satellite galaxies orbiting the Milky Way. Among the more than 1700 variable stars she identified there, she found a number of what we now call Cepheids. In the Small Magellanic Cloud she was able to plot light curves for

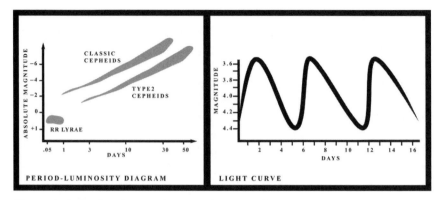

Figure 1.5 Cepheid Variables and the Period-luminosity Relationship

twenty-five Cepheids, and noted that the brighter the star appeared, the longer its period of pulsation. As she explained in her 1912 paper,

> there is a simple relation between the brightness of the variables and their periods. . . . Since the variables are probably at nearly the same distance from the Earth, their periods are apparently associated with their actual emission of light, as determined by their mass, density, and surface brightness. (Leavitt 1912, 3)

In other words, as seen in Figure 1.5, period is related to the absolute magnitude of the stars. This *period-luminosity relationship* is one of the most powerful relations in astronomy, and after her death in 1921, she was nominated for the Nobel Prize in physics by an astronomical colleague who was unaware of her demise. For a random Cepheid, once its period is measured, its absolute magnitude can be estimated, and from that the distance calculated (through the distance modulus formula). The only problem is that the exact numerical relationship between absolute magnitude and period must be first determined using Cepheids of known distance.

The first person to attempt such a calibration was Ejnar Hertzsprung, who was also the first scientist to use the generic label "Cepheid" to describe these stars. In 1913 he used the observed motions of thirteen Cepheids in the Milky Way too distant for parallax measurements as a statistical way to estimate their distances. From this he determined a mathematical expression for the period-luminosity relation and used it to estimate the distance to the Small Magellanic Cloud. Although his calibration wasn't correct, it was an important first step. As we shall see, the period-luminosity relationship is now the most reliable means of determining the distances to galaxies, so long as individual Cepheids can be viewed. The first widely used calibration was done by Harlow Shapley, later the director of the HCO. Unbeknownst to Shapley and the astronomical community, it contained several errors, and resulted in erroneous calculations of distances to galaxies and the size of the Milky Way for several decades.

The most fundamental error in Shapley's calibration was discovered in the 1940s by Walter Baade. Using the largest telescopes in the world, the 100-inch at Mount Wilson and later the 200-inch at Mount Palomar, he studied individual stars in the Andromeda Galaxy (M31). He found that the spiral arms had more O and B stars, while the nucleus had more red giants, and named these two types of stars *Population I* and *Population II*, respectively. He also noted that the Population I stars were similar to stars in the spiral arms of the Milky Way, while the Population II stars were reminiscent of stars in globular clusters. Baade had stumbled upon the same division in stellar chemistry as Nancy Roman had in her luminosity class studies, namely that stars of different compositions have different intrinsic brightnesses and occupy slightly different locations on the HR diagram, even for the same spectral class. For example, main sequence stars are population I stars, while normal stars of population II occupy a parallel but slightly lower line on the HR diagram, the subdwarfs. In discussing stellar chemistry, astronomers use the term *metallicity*, which compares the mass of elements heavier than helium to the overall mass of the star. For example, the metallicity of the sun is 0.02, so 2 percent of the sun's mass is in elements heavier than helium.

Baade found 300 Cepheids in M31 and found that they come in Population I and Population II varieties, with slightly different shapes to their light curves, and importantly, different period-luminosity relations. Type I Cepheids, called classical Cepheids, are Population I stars and are about 1.5 magnitudes brighter than type II (Population II) Cepheids (now called W Virginis stars). By unknowingly lumping together both types of Cepheids into his calibration of the period-luminosity relation-ship, Shapley introduced significant error into his distance calculations. As with many astronomical concepts we have encountered thus far, the naming system of the populations was done before the physics behind the differences was understood. Population II stars have lower metal-licities because they represent an earlier generation of stars, and were made of material that had not been recycled and enriched as much as that contained within more recent generations of stars (Population I). Extending this logic, there should exist a first generation of stars, referred to as *Population III* in keeping with the history of the naming of populations, which contains no enriched material. This primordial generation of stars would have been comprised of pure hydrogen and helium, and as we shall see, understanding them presents unique challenges to astronomers.

Stellar Corpses

Once a star reaches the end of its fusion cycles, its final state of rest depends on its mass. Low mass stars collapse to around the size of the earth, forming a white dwarf. Further collapse is prevented by the degeneracy pressure of the electrons. As we have noted, the degeneracy pressure is created by the electrons being forced as close together as nature will allow, having filled up all available energy states. This is a result of the *Pauli exclusion principle,* one of the many predictions of quantum mechanics. Simply put, no two electrons can occupy the same quantum state, just as no two students are allowed to occupy the same seat in a classroom. Gravity can cram together the electrons' available states, just as the seats in a classroom can be shoved closer together, but there comes a point where neither can be moved closer together. This halts the gravitational collapse of the star and holds up the white dwarf from further collapse. The strength of the electron degeneracy pressure can only counter gravity if the star dies with a mass under 1.4 solar masses, named the Chandrasekhar limit after Nobel Prize laureate Subrahmanyan Chandrasekhar who first calculated it. Stars that die with greater masses explode as a supernova. Note that it is the star's final mass, not initial mass, that matters. There are various ways that a star can shed mass during its final stages of life, such as a planetary nebula.

There are actually two main classes of supernovae, which explode for different reasons. The first, type Ia, involve a white dwarf and its red giant binary partner. Mass will flow from the bloated red giant, adding material to the white dwarf. Since white dwarfs are on a strict weight regiment, bound by the Chandrasekhar limit, they will resist accumulating too much mass. As the material spirals onto the white dwarf, it first forms an *accretion disk.* If enough material accumulates, it can ignite, blowing off sufficient mass to avoid reaching the critical limit. This temporarily increases the apparent brightness of the system, creating a *nova.* If too much material accumulates and the Chandrasekhar limit is reached, the carbon and oxygen in the white dwarf ignite in a sudden burst of fusion, and the star explodes, creating a *type Ia supernova* (SNIa). In a *type II supernova* (SNII), a red supergiant explodes once it has created an iron core and fusion has suddenly ceased. Note that what

Tycho Brahe originally named a nova was actually a supernova. The difference between the two was not understood until the early twentieth century.

After a type II supernova explosion, the remnant of the core—if any—collapses to a diameter of about 20 km. Since its mass is above the Chandrasekhar mass limit, the degeneracy pressure of the electrons is of no use. Rather than violate the Pauli exclusion principle and "sit in each other's laps," the electrons combine with protons to form neutrons, with a wind of neutrinos traveling near the speed of light screaming out of the dying star hours before the explosion is seen at the star's surface. This type of stellar corpse is therefore composed of neutrons, and their degeneracy pressure prevents further collapse once the corpse has compressed to the size of a large city. The first neutron stars were discovered by British graduate student Jocelyn Bell in 1967 as part of a radio survey of quasars. She found unnaturally regular and rapid radio signals emanating from some unidentified celestial sources, which were briefly dubbed LGM (little green men). These *pulsars* are now known to be neutron stars rotating many times a second. Electrons spiraling in their intense magnetic fields emit a characteristic type of radio waves known as *synchrotron radiation* because it was first seen in atomic collision machines. These radio waves form two beams, at the north and south magnetic poles of the neutron star, and as it rotates, beams of radio waves sweep past Earth like a lighthouse beacon.

Neutron degeneracy pressure can prevent complete gravitational collapse only if the mass is under about 2.5 solar masses. For dying stars heavier than this, there is no force in the universe that can halt the collapse, and the material implodes to a mathematical point called a *singularity*, where the laws of physics, including general relativity, break down. Such an object's gravitational field is so strong (or in relativistic language, it warps space-time to such a high degree) that not even light can escape. Such an object was named a *black hole* by physicist John Wheeler in the 1960s. Because of its very nature, a black hole is not directly observed. Rather, its presence is discovered by its gravitational effect on nearby material. For example, Cygnus X-1, the most famous black hole, was discovered because of its interaction with its blue giant companion star HDE226868 (the 226,868th star Annie Cannon classified as part of the Henry Draper Catalog extension). As material is drawn off the normal star, it spirals into an accretion disk before plunging down the gravitational drain into the black hole. The particles are moving so fast that collisions heat the disk to a million Kelvin, creating x-rays, which in turn led to the discovery of the system by the Uhuru x-ray satellite in the 1970s. Black holes of stellar size (typically less than 20 solar masses) are certainly fascinating creatures, but nature creates monstrous black holes of millions of solar masses, which lurk in the hearts of galaxies. It is these which will occupy some of our attention in a future section.

2

THE HISTORY OF COSMOLOGY

THE RISE OF THE GEOCENTRIC UNIVERSE

Cultures throughout the world developed their own creation myths, and over time many were able to utilize careful observations to discern the natural cycles of eclipses and apparitions of the planet Venus as the Morning and Evening Star. Modern cosmology traces its history as a continuous tradition of thought from the ancient Greek philosophers (circa 600 B.C.) through the Renaissance to Isaac Newton. As with many other vigilant observers of the world, the scholars of ancient Greece understood that the earth was round. The fact that different stars were visible as one moved closer to the North Pole, the observed sinking of ships below the horizon as they sailed out from the coast, and the round shadow Earth cast upon the moon during a lunar eclipse led them to such knowledge. This may have first been articulated by the famed mathematician Pythagoras of Samos (b. circa 572 B.C.). Later Greek historians attributed to him the understanding that the Morning Star and Evening Star were both the same body. Pythagoras pictured the heavenly bodies as being perfect spheres, moving in equally perfect circles. As the sun, moon, and planets moved, they created an ethereal music beyond the profane ear of mere mortals to discern, the famous "harmony of the spheres."

When asked what he felt was the reason for being born, Anaxagoras (b. circa 500 B.C.) is reported to have replied, "To investigate the sun, moon, and heaven" (Heath 1991, xxxiii). He understood that the moon shines by reflected sunlight, and with this knowledge correctly described the causes of both solar and lunar eclipses. He is also attributed with the concept that the stars are distant fires, but he felt they were cooler than the sun because they were found in a colder environment. Eudoxus (b. 408 B.C.), a student of Plato, developed a *geocentric*, or Earth-centered, model for the universe. Based on thirty-three rotating spheres, the model sought to explain the apparent motions of the sun and moon with three spheres each, while the five known planets (Mercury, Venus, Mars, Jupiter, and Saturn) were said to have four spheres of motion each.

As complex as Eudoxus' model may seem, an even more intricate model of fifty-five concentric crystalline spheres was developed by famed philosopher Aristotle (b. 384 B.C.). All heavenly bodies moved around the earth, with the stars located on another, fixed crystal sphere, located some large distance beyond the orbit of Saturn. This celestial sphere was spun by the "Prime Mover," a mystical force that was interpreted by later Christian writers as angels or even the hand of God. Aristotle argued that the universe was perfect, unchanging, and eternal, without beginning or end. It was this eternal nature of the heavens which formed part of his argument for an Earth-centered universe. Aristotle offered that if the earth moved, it would be "unnatural" and could not be sustained, since objects seem to fall toward the center of the earth. He also argued that Earth had to move toward the center of the universe, but since everything appears to move toward Earth's center, it must be the center of the universe. He offered that if Earth were to orbit the sun or some other body, the positions of the stars relative to each other would seem to shift as we observe the stars from slightly different perspectives in that orbit. Since this *parallax* had not been observed, Aristotle argued that this proved that the earth could not move. Finally, Aristotle appealed to everyday observations—we don't fly off Earth, or even sense an overall motion, so it appeared to him and many of his contemporaries that Earth simply could not be moving.

Just as Aristotle noted that objects seemed to fall toward the center of the earth, he saw fire rise toward the heavens, and concluded from this that Earth and the heavens must be made of two completely different materials. Since it was commonly believed that Earth was made of four basic elements (fire, earth, air, water), he proposed a fifth element, called aether, which is perfect and unchanging, like the heavens. The direct influence of Aristotle's writings can be traced through European science until openly challenged by the observations of Galileo in the seventeenth century.

A completely different view on the heavens was expounded by Aristarchus of Samos (b. circa 310 B.C.). Using the geometry of lunar eclipses and the phases of the moon, he argued that the moon is about 36 percent the diameter of the earth, the diameter of the sun is about seven times larger than the diameter of the earth, and that the distance between Earth and the sun is about 19 times larger than the distance between the earth and moon. His calculations were grossly inaccurate, not surprising given they were based on naked-eye observations. The actual values are approximately 27 percent, 109, and 388, respectively. Despite these errors, he was successful in showing that the sun is much larger than the earth, and advocated for a *heliocentric*, or sun-centered, model for the universe.

Perhaps the most famous experiment of ancient times was Eratosthenes' determination of the circumference of the earth. The director of the famous library of Alexandria, Eratosthenes (b. circa 276 B.C.) learned that on the first day of summer, the sun passed directly overhead at Syene (now called Aswan). He watched the sun at Alexandria on the summer solstice, and found that it did not pass directly overhead, but instead passed 7 degrees lower (a fact he measured by observing the shadow cast by an upright stick). Not only was this difference caused by the curvature of the earth, but because the earth is roughly a sphere, by knowing the difference in angle between the two cities and the distance between them, one can calculate the circumference of the earth. Since 7 degrees is about 1/50 of the 360 degrees in a circle, the distance between the two cities must be 1/50 of the total circumference. The answer he derived, approximately 25,000 miles, is very close to the actual value of 24, 902 miles. The largest source of error is the fact that the earth is not a perfect sphere.

Among the greatest astronomers of antiquity was undoubtedly Hipparcos of Nicaea (now Iznik, Turkey). Between 150 and 125 B.C., he was responsible for an impressive number of observations, inventions, and theoretical models. The magnitude system previously described is attributed to him, as is the discovery that Earth wobbles on its axis like a top, causing the position of the sun relative to the stars to shift over the centuries. This precession now places the sun in the constellation of Pisces on the first day of spring, or vernal equinox, soon to shift into Aquarius (as lauded in the famed song by the Fifth Dimension). He is frequently credited with the invention of trigonometry and the astrolabe, a device used to measure the positions of the stars.

The work of Hipparcos was very influential on Ptolemy, whose *Almagest* was considered a seminal work in astronomy through the sixteenth century. Working in Alexandria, Egypt between A.D. 127 and 141, Ptolemy wrote extensively on a detailed geocentric model of the cosmos. He based his model on the postulates that

The heavens are spherical and move spherically; that the earth, in figure, is sensibly spherical also when taken as a whole; in position, lies right in the middle of the heavens, like a geometrical center; in magnitude and distance, has the ratio of a point with respect to the sphere of the fixed stars, having itself no local motion at all. (Munitz 1957, 106)

This was a classically Aristotelian view, holding that the heavens were perfect and unchanging (or "immutable"). However, when Ptolemy tried to tackle the very practical problem of trying to model the apparent motions of the sun, moon, and planets, like other geocentric thinkers before him he found that simple circular orbits centered on Earth were insufficient. He expanded an early model created by Hipparcos, and placed each planet on a small circle called an *epicycle*, which, as shown in Figure 2.1, was then attached to a larger circle, called the *deferent*. Unlike Hipparcos, Ptolemy then added a further fudge factor, stating that the earth is not exactly at the center of planetary motion, but that instead the true center of motion was a point called the *equant*, which like Earth was offset from the center of the circular deferent. By carefully adjusting the epicycle, deferent, and equant for each planet, Ptolemy was able to model the observed motions of the planets in the sky. Although their apparent motion was nonuniform (sometimes moving faster or slower relative to the background stars than on average), as seen from the equant the motion would be perfectly uniform. This prejudice in demanding the perfection of the heavens was to prove to be the bane of astronomy for centuries to come.

Over the next few centuries, errors crept into the predicted positions of the planets, but these were fixed by adding additional epicycles to the model. Thus observational accuracy was achieved at the expense of elegance and simplicity. With the fall of the Roman Empire the works of Aristotle and Ptolemy were largely lost in Europe, but were saved by the Islamic world, which achieved its Enlightenment between the eighth and tenth centuries. In fact, the name now used for Ptolemy's work comes from an Arabic term meaning "greatest." When the Dark Ages ended in Europe and the great works of the past were rediscovered, the geocentric models of Aristotle and Ptolemy were embraced by the Catholic Church as being in line with Christian doctrine. It was this alignment with the powerful religious institutions of Europe which made the overthrow of the erroneous geocentric tradition a lengthy process—namely the Copernican Revolution.

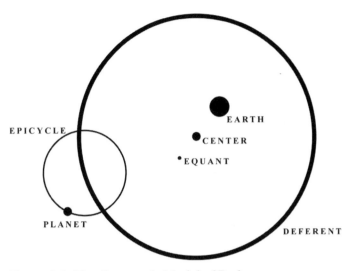

Figure 2.1 The Geocentric Model of Ptolemy

THE COPERNICAN REVOLUTION

Nicolaus Copernicus (1473–1543) was a canon in the Catholic Church in Poland, and even earned a degree in canon law. Based on observations of the night sky and the continuous errors which had crept into Ptolemaic predictions of planetary positions over the centuries, he proposed a heliocentric model of the heavens. As laid out in his privately circulated, brief work, *Commentariolus* (circa 1514), the Copernican model was based on several axioms:

1. Stars appear to rise and set because of the rotation of the earth.
2. Earth, like all the planets, revolves around the sun in a circular orbit.
3. The apparent backward motion shown by some planets—called retrograde motion—is an optical illusion caused by the faster moving Earth passing out the slower moving outer planets, such as Mars.
4. The lack of observable parallax is expected because the stars are simply too far away.

A much longer treatise, *De Revolutionibus Orbium Caelestium* (On the Revolutions of the Heavenly Spheres), remained unpublished until just before his death. Georg Joachim Rheticus, a former student, took charge of the publication, but the final details fell to Lutheran theologian Andreas Osiander. In a controversial, unsigned preface, Osiander attempted to appease the majority of readers, who would generally ascribe to the geocentric universe (partially for religious reasons). He wrote,

> it is not necessary that these hypotheses should be true, or even probable; but it is enough if they provide a calculus which fits the observations. . . . Let no one expect anything in the way of certainty from astronomy, since astronomy can offer us nothing certain, lest, if anyone take as true that which has been constructed for another use, he go away from this discipline a bigger fool than when he came to it. (Hawking 2002, 7–8)

Osiander's preface appears even more an affront when compared to the preface Copernicus himself had meant to include, which defiantly accosted possible detractors:

> Perhaps there will be babblers who, although completely ignorant of mathematics, nevertheless take it upon themselves to pass judgment on mathematical questions and, badly distorting some passages of Scripture to their purpose, will dare to find fault with my undertaking and censure it. I disregard them even to the extent of despising their criticism as unfounded. (Singh 2004, 43)

As Copernicus had predicted, there were those who attacked his model on religious grounds. It was feared that a heliocentric universe would remove humans from a place of centrality (as God's special creation) and lead to the heretical view that humans were merely one part of the natural world. Martin Luther himself attacked the Copernican model, remarking that "The fool wants to turn the whole art of astronomy upside-down" (Singh 2004, 40).

For religious as well as scientific reasons, the Copernican model was slow to be accepted. Not only was the idea of a moving Earth hard to swallow for many people, but the model didn't predict the motions of the planets with any better precision than the geocentric model. In fact, Copernicus had to utilize small epicycles in order to achieve a reasonable match with the observed heavens. The basic problem was that Copernicus had fallen prey to the same prejudice which Aristotle had held supreme—namely the perfection of the heavens and reliance on circular motion. Copernicus wrote that "we must observe that the Universe is spherical. This is either because that figure is the most perfect . . . or again because all the perfect parts of it, namely, Sun, Moon and Stars, are so formed" (Munitz 1957, 152–153). So long as astronomy was a slave to this prejudice, attempts to discern the true motions of the planets were doomed to failure. But Copernicus' model was an improvement over the geocentric model of Ptolemy and his successors in other ways which were not immediately appreciated. Not only did it explain the lack of observed parallax and provided a natural explanation for retrograde motion (as opposed to the geocentric practice of tweaking the various circles to mimic the effect), but it allowed for the calculation of the distances of the planets in order from the sun, at least in units of Earth's orbital size (the astronomical unit).

KEPLER'S LAWS OF PLANETARY MOTION

One famous observational astronomer who was troubled by Copernicus' model was Tycho Brahe (1546–1601). The Danish astronomer was kidnapped by his uncle while still a toddler, under the excuse that Tycho's father already had two sons. This proved fortuitous for Tycho, who received the best schooling available to nobility. He became interested in observational astronomy as a teenager, after viewing a solar eclipse. Unfortunately, less positive pursuits also engaged his time, and in 1566 he lost the tip of his nose in a sword duel. The wound never healed properly, and Tycho took to wearing a metal replacement for the rest of this life.

Tycho achieved fame through his meticulous observations of a "new star," which he dubbed a *nova*, in 1572. By observing the position of the object relative to the other stars as seen from several locations, he found that like the distant stars it did not seem to shift, or exhibit parallax. But the moon, as seen simultaneously from different locations, does show a shift in position relative to the stars. Tycho had therefore shown that this nova was far beyond the moon, and must, in fact, be among the known stars. The appearance of this previously unseen star seemed to fly in the face of the Aristotelian idea of the perfection and immutability of the heavens, but its existence could not be denied, as it could be seen by anyone bothering to gaze skyward.

King Frederick II of Denmark gave Tycho the island of Hven in 1576, where the astronomer built his observatory, Uraniborg, equipping it with the finest naked-eye instruments for measuring the stars. For 20 years, Tycho made precise measurements of the motions of the planets. He came to appreciate some aspects of the Copernican model, but took issue with the concept of a moving Earth as well as its seeming contradiction with a literal translation of Scripture. In response, Tycho proposed a compromise model in 1588, which had the sun, moon, and stars orbiting Earth, but the planets orbiting the sun. This Tychonic hybrid model achieved some level of popularity for several decades.

Following the death of his patron, Tycho received a position in Prague as Imperial Mathematician in the court of Emperor Rudolph II in 1599. It was here that he took Johannes Kepler (1571–1630) as an assistant. Kepler, a mathematics teacher in Graz, had come to the attention of Tycho as the author of *Mysterium Cosmographicum.* A year later, Tycho died, reportedly as the result of a drinking incident. Kepler succeeded his mentor in the court, and by his own admission, "quickly took advantage of the absence, or lack of circumspection, of the heirs, by taking the observations under my care, or perhaps usurping them" (Hawking 2002, 631). Tycho's legacy of observations afterward became the basis of Kepler's famous *Rudolphine Tables* of planetary positions.

Kepler's main project, begun at the suggestion of Tycho, was an attempt to explain the peculiarities in the observed motion of Mars. Like astronomers before him, he was handicapped for years by a conviction that the heavens must be based on perfect, uniform, circular motion. It was only after attempting a noncircular shape, namely an ellipse, that he cracked the problem wide open. The result was Kepler's Three Laws of Planetary Motion, which have ostensibly not changed since their original publication:

1. The orbit of a planet (including Earth) is an ellipse, with the sun at one focus. The other focus is a point of empty space. For those not familiar with the geometry of an ellipse, the two foci are points on either side of the center, along the wide side of the ellipse.

2. A line connecting a planet to the sun sweeps out an equal area of space in an equal amount of time at any point in its orbit. As a result, a planet moves fastest along its orbit when nearest the sun, and slowest along its orbit when farthest from the sun. Note that this clearly contradicts the Aristotelian idea of uniform motion.

3. There is a simple relationship between the period of the orbit (P) and a planet's average distance from the sun (a), such that $P^2 = a^3$, where P is measured in years and a in au.

The first two laws appeared in *The New Astronomy* (1609), while the third appeared later, in *The Harmony of the Spheres* (1619). Kepler's laws were empirical; that is, strictly based on observations. Not only did they explain Tycho's decades of observations of Mars, but they could predict the future motions of the planets with greater precision than any previous model. The one obvious weakness in Kepler's model was a lack of theoretical underpinning. What was the force that kept the planets orbiting the sun? Gravity, as the force we understand today, had not been conceptualized at this time. Kepler proposed that the sun emitted a type of magnetic force which interacted with the magnetic fields of the planets to produce the orbits. Therefore, while Kepler had come upon essentially the correct model of planetary motion, it took further advances in theory, observation, and most importantly, technology, to convince the astronomical community.

GALILEO, NEWTON, AND THE TELESCOPIC REVOLUTION

As previously noted, the first astronomer to use a telescope in an observing program was Galileo Galilei (1564–1642). The Italian professor of mathematics and physics first used a crude refractor made of two eyeglass lenses. He later learned how to grind his own lenses, and succeeded in making an instrument which magnified twenty times. Among his important observations were craters and mountains on the moon and sunspots, both of which contradict the ancient idea of the perfection of the heavens. By monitoring sunspots, an experiment which must be done with extreme care to prevent blindness, he showed that the sun rotated like Earth, only much more slowly (about once a month). He was able to resolve the Milky Way into individual stars, and found that the planets were not point-like as the stars were, but instead appeared as small disks. This agreed with the prediction of Copernicus that the stars were much farther than the planets.

Two additional observations greatly contributed to Galileo embracing the Copernican concept. Firstly, he found that the planet Venus showed a full range of phases like the moon, an observation which cannot be explained by a geocentric model, no matter the amount of tweaking. Secondly, he discovered four moons orbiting Jupiter, which demonstrated that the commonly held belief that if Earth moved, the moon would fly off into space, was clearly wrong. It also proved that Earth was not the center of all motion in the universe. His discoveries were published in a pamphlet entitled *Sidereus Nuncius* (Starry Messenger) in 1610.

Galileo was fully aware of the religious concerns about removing Earth from the center of the universe. Several years after *Sidereus Nuncius*, he circulated a series of letters that argued against a literal interpretation of the Bible when it came to astronomical topics. Soon after, in 1616, the Vatican issued an edict against the teaching of the Copernican doctrine. Galileo openly defied that edict in 1632 with the publication of *Dialogues Concerning the Two Chief World Systems*. Written in Italian rather than Latin, it was widely read, and its ridicule of the Aristotelian universe and open support for Copernicus set him on a collision course with the Catholic Church. In 1633 he was put on trial by the Inquisition for heresy, and was found guilty. He was forced to recant his support of the Copernican doctrine, and spent the rest of his life under house arrest. In 1979, Pope John Paul II called for a commission to reevaluate Galileo's trial, with the result that in 1983 it was announced that Galileo should not have been condemned. The commission's report was officially accepted by John Paul II in 1992. Galileo is generally considered to be the last astronomer to be harassed for espousing that Earth orbits the sun.

The same year as Galileo's death, Isaac Newton (1642–1727) was born in Lincolnshire, England. He spent 4 years as a student at Trinity College, but returned to his family home when the university closed in 1665 for 2 years due to the plague. Temporarily unable to continue with formal studies, Newton set to experimentation on his own, later resulting in many important discoveries. Among his impressive list of contributions to science and mathematics are the reflecting telescope, studies of the spectrum of light, and calculus. His most famous written work, *Philosophiae Naturalis Principia Mathematica* (The Mathematical Principles of Natural Philosophy), or *Principia*, as it is commonly known, was not published until 1686. Newton was generally miserly when it came to sharing his discoveries with the wider scientific community, and *Principia* was produced through the insistence of Edmund Halley, who had used Newton's techniques to determine the orbit of the comet that now bears his name.

The first of the two most widely known topics included in the *Principia* is Newton's Three Laws of Motion, which still form the basis of the study of mechanics and motion, and deservingly acquire their own chapter in most introductory physics textbooks:

1. A body in motion remains in motion, and a body at rest remains so, unless acted upon by an outside force.
2. The force required to change the motion is proportional to both the mass of the object, as well as the size of the change in motion (or acceleration). In mathematical form, $F = ma$.
3. Every action is accompanied by an equal and opposite reaction.

The second vital topic is the law of universal gravitation, which states that all objects in the universe attract all other objects through a force that decreases in strength as the distance between them increases via an inverse-square relation (just as in the case of the intensity of light).

If the moon were suddenly moved twice as far as it currently orbits, the gravitational force between it and Earth would drop to one-quarter its current value. The gravitational force also directly depends on the masses of the objects in question. If the moon's mass were to suddenly double, the gravitational force between the earth and moon would also double. In fact, using the law of universal gravitation, the three laws of motion, and calculus, Newton was able to derive Kepler's laws from first principle, an exercise routinely repeated by undergraduate physics majors to this day.

With the law of universal gravitation, Newton completed the Copernican Revolution. The motions of the planets were understood to be caused by laws of nature, and the heavens worked by the same mundane rules that governed falling bodies on Earth. But the theological tension surrounding cosmology had not ended. Newton himself held unique religious beliefs which had their root in Unitarianism. He wrote in the *Principia* that there existed "absolute, true, and mathematical time, of itself and from its own nature, [that] flows equably without relation to anything external, and by another name is called 'duration'. . . . Absolute space, in its own nature, without relation to any thing external, remains always similar and immovable" (Munitz 1957, 202). Absolute time and space were intimately connected with God in Newton's mind. In an unpublished manuscript, he wrote that an "infinite and eternal divine power . . . extends infinitely in all directions" (Harrison 2000, 56). In response to the idea that gravity could not act through empty space, he argued that empty space is actually filled with spirit, and those who denied the possibility of empty space actually denied the existence of God.

Newton's thoughts concerning the universe as a whole were also shaped by a series of letters exchanged with Richard Bentley, a clergyman. When asked what would happen if a finite universe were filled with evenly spread matter, Newton answered that it would naturally collapse into a single spherical mass under the influence of gravity, but that if space were instead infinite, the matter would clump together into individual masses, like stars in the observed universe. This suggested stability fit with Newton's concept of a static and unchanging universe. He believed that if there was a large-scale motion there would have to be a center of that motion, something he disavowed. Unfortunately, Newton was wrong on both accounts. An infinite, static universe is an impractical balancing act, and is inherently unstable. Secondly, large-scale motion does not require a physical center. Despite this instability being pointed out by Bentley, Newton remained convinced that the universe was static and infinite, in a state of perfection reflecting his concept of the divine. In fact, he believed that it was the hand of God which kept the universe in balance. This deep-rooted belief in a static universe was to plague cosmology through the beginning of the twentieth century, much as the perfection of the heavens had done so in the previous millennium.

WHY IS THE NIGHT SKY DARK?

One of the ramifications of an infinite, eternal, static universe was known before Newton's time—namely an apparent paradox. If there were infinite stars spread throughout infinite, eternal space, then anywhere one would look in the sky, one's eye should intersect with the light from some distant star. Therefore the night sky should not be dark, but instead have an overall glow. This was not observed, hence the paradox. Perhaps the first clear articulation of this problem was given by Johannes Kepler, in his pamphlet written as a response to Galileo's work, called *Conversation with the Starry Messenger.*

You do not hesitate to declare that there are visible over 10,000 stars. The more there are, and the more crowded they are, the stronger becomes my argument against the infinity of the universe.... [I]f they are suns having the same nature as our sun, why do not these suns collectively outdistance our sun in brilliance? Why do they all together transmit so dim a light to the most accessible places? (Harrison 1987, 50)

Kepler was using this paradox as an argument against the infinity of the heavens. This was again articulated in *Epitome of Copernican Astronomy* (1618), his final major work, where he offered that the finite heavens were "enclosed and circumscribed as by a wall or a vault" (Ibid.). A century later, Edmund Halley attempted to solve the problem of the dark sky by claiming that most stars in the infinite universe are simply not visible because their light is too feeble to reach us. Therefore, he erroneously believed, we would see a dark sky.

A more correct analysis was done in 1744 by Swiss astronomer Jean-Philippe de Chéseaux, who believed that the solution was that most of the distant starlight was absorbed by intervening clouds in space. This solution was offered again in 1823 by German astronomer Heinrich Wilhelm Olbers, in his article "On the Transparency of Space." Although he discussed the problems with Kepler's argument, he did not mention de Chéseaux's analysis, and in fact came to an identical solution. It is because of the widespread readership of Olbers' article, and his faulty citation of previous literature, that the problem became known as *Olbers'* *paradox*. Although this solution became popular, it was easily proven ineffective. John Herschel wrote in *The Edinburgh Review* in 1848 that if interstellar material absorbed starlight, it would eventually become hot itself and then radiate light "from every point at every instant as much heat as it receives" (Harrison 1987, 99). This critique was reinforced by William Thomson (Lord Kelvin), several decades later.

The correct explanation was actually proposed by German astronomer Johann Heinrich von Mädler in 1861. If one gives up the notion of an eternal universe, then there would have only been a finite time for light to reach our eyes. For example, if the universe were 10 billion years old, then only the light from stars within a shell of radius 10 billion light years in any direction could have reached us, no matter the actual size of the entire universe. This limiting distance is called our horizon. While this certainly contains a large number of stars, that number is not infinite, therefore one can show mathematically that the night sky would appear dark. Although von Mädler had essentially solved the problem, compared to the astronomical understanding of his day, his proposal was nothing more than speculation. A theoretical basis for his suggestion would not come for several decades.

THE PROBLEM OF THE NEBULAE

Ancient cultures stared at the hazy band of light arching from horizon to horizon and explained it as the backbone of the night, or spilled milk from a goddess' breast. Modern sky watchers know it as the Milky Way, our home galaxy, one of millions now known in the wider universe. The road to our current perspective has not been a smooth one. The first detailed model of the Milky Way was proposed by English astronomer Thomas Wright in his 1750 book *An Original Theory of the Universe*. He pictured the Milky Way as a huge disk or sphere of stars, which includes the sun. These stars orbited the center of the Milky Way—what he termed a "center of creation"—like planets orbiting the sun. Most importantly, he claimed that "many cloudy spots just perceivable by us," increasingly seen in ever-more powerful telescopes—the *nebulae*—"may be external creations, bordering upon the known one, too remote for even our telescopes to reach" (Macpherson 1919, 329).

Immanuel Kant, the famed German philosopher and scientist, read a review of Wright's book and was greatly influenced by it. He laid out his cosmological viewpoint in *Universal Natural History and Theory of the Heavens* (1755). Gravity was the glue that held the universe together, as the mutual gravitational attraction of the stars for each other perfectly balanced their individual motions, preventing chaos from ensuing. Like Wright, he pictured the Milky Way as a rotating disk of stars. He also held that the nebulae were other Milky Ways, seen at large distances. Kant also argued for an infinite universe, largely on the theological grounds that limiting space effectively limited the grandeur of God:

We come no nearer the infinitude of the creative power of God, if we enclose the space of its revelation within a sphere described with the radius of the Milky Way, than if we were to limit it to a ball an inch in diameter. . . . Eternity is not sufficient to embrace the manifestations of the Supreme Being, if it is not combined with the infinitude of space. (Kragh 1996, 3)

The word nebula comes from the Latin term for a cloud, and until several decades into the twentieth century it was used as a generic term for any object that had a fuzzy appearance when viewed through a telescope at low power. This includes true clouds of gas and dust (what we today still call a nebula), as well as clusters of stars and galaxies, both of which can be resolved into their component stars with powerful enough instruments. In the eighteenth and nineteenth centuries, increasing numbers of these objects were discovered, classified, and cataloged. The true nature of these bodies became a subject of considerable debate, especially those which had a spiral structure, or *spiral nebulae.* Today we know these objects as spiral galaxies, such as the Milky Way and the Andromeda Galaxy.

The first large catalog of nebulae was created by Charles Messier, the famed French comet hunter and observing assistant at the Marine Observatory in Paris. In the course of sweeping the sky looking for comets (named from the Greek term for fuzzy or hairy stars), he would occasionally run across other fuzzy objects which one might confuse for a comet. He kept a running record of these objects and their locations in the sky, and published a list of 68 such objects in 1780. A later edition included 103 objects, which today comprise the Messier Catalog. For example, the Orion Nebula is M42 and the Andromeda Galaxy (then called the Andromeda Nebula) is M31. The catalog is sometimes given as including 110 objects, because several objects which Messier probably knew about were posthumously added to the catalog. Since these objects are all visible with an 8-inch telescope from dark skies, they comprise the foundation for most amateur astronomers' observing programs.

The number of known nebulae soon mushroomed due to the work of William Herschel. The German musician immigrated to England in 1757 as a young man to escape the Seven Years War. He became interested in astronomy and began building his own reflecting telescopes, eventually inventing a type of instrument which, unlike Newton's original reflector, did not use a secondary mirror. His sister Caroline joined him in England 15 years later and became his observing and telescope-making assistant and a comet hunter in her own right. The largest instrument they constructed was completed in 1788, a behemoth 48-inch aperture reflector with a 40-ft focal length, which remained the largest telescope in the world for nearly 60 years. Herschel became well known in 1781 when he discovered the planet now called Uranus. He was appointed court astronomer by King George III (after whom he originally tried to name the planet) and his astronomical career blossomed.

Two of Herschel's observing projects were observing nebulae, and mapping the structure of the Milky Way through a technique known as star gauging. This involved counting the number of stars visible in a standard size region of the sky (such as that visible through a paper towel tube held up to your eye) in different directions. Assuming (incorrectly) that all stars have the same intrinsic brightness or absolute magnitude, those that appeared dimmer were assumed to be more distant. The result was an influential model of the Milky Way that resembled a disk or "grindstone," with the sun predictably located at the center. He originally proposed a diameter of 1,800 pc (1785), which he increased to over 6,000 pc by 1806 (Trimble 1995, 1139). One could say that the Copernican Revolution took a slight step backward, because even though Earth wasn't the center of the solar system, our solar system was believed to be the center of the known universe.

In terms of the nebulae, Herschel's observations also proved of lasting importance. In a 1784 paper he reported the discovery of about 500 new nebulae, and also announced that he had been able to resolve most of the objects in the Messier catalog into individual stars. This suggested that these nebulae were actually star systems located at great distances, and some could even "outvie our Milky Way in grandeur" (Macpherson 1919, 329–330). The term *island universe* was coined to describe such large extragalactic systems. By 1786 he had published a catalog of 1,000 nebulae, followed by a second catalog of 1,000 more nebulae in 1789, and a final catalog of 500 additional objects in 1802. Herschel's son, John, became an astronomer of note, and published a cumulative catalog of over 5,000 nebulae and star clusters in 1863. As William observed more nebulae, he found that there were some which could not be resolved into individual stars, even with his large telescope. He began to question the *island universe hypothesis*, and by 1811 seems to have decided that most of the nebulae were in fact objects within the Milky Way.

William Parsons, more commonly known as Lord Rosse, took telescope making to another level in 1845 with the completion of a 72-inch aperture reflector with a 54-ft focal length. The so-called "Leviathan of Parsonstown" was build on the grounds of Lord Rosse's family castle in Ireland, and remained the largest telescope in the world until 1917, when the 100-inch Hooker telescope opened at Mount Wilson in California. One of the first discoveries made with the Leviathan was the spiral structure of M51, the Whirlpool Nebula. Within five years, Lord Rosse reported the discovery of 14 other spiral nebulae, although some of these are now known to be spurious. Around this time, the island universe hypothesis enjoyed renewed support, partly through *Kosmos*, the popular 1845 work by famed German geographer Alexander von Humboldt. In addition, Lord Rosse was able to resolve even more nebulae into individual stars, leading John Herschel to admit in 1849 that it might be theoretically possible to resolve all nebulae.

Around this time, spiral nebulae became the focus of cosmological debate and study. In 1852, Stephen Alexander, a Professor of Astronomy and Mathematics at what became Princeton University, published a lengthy treatise on the classification of nebulae. He claimed that the Milky Way is, in fact, a spiral nebula, with up to four spiral arms. If the spiral nebulae were similar to the Milky Way in shape, then it seemed logical for them to indeed be separate "Milky Ways" outside of our galaxy. Another important fact which set spiral nebulae apart from other types of nebulae and star clusters was their location in the sky. They tended not to be found within the plane of the Milky Way (called the *zone of avoidance*), but instead were numerous when looking out of the disk of the galaxy, toward the galactic poles. The preponderance of evidence seemed to point in the favor of the island universe hypothesis, and for a time it gained increasing support in the astronomical community.

But dissenting voices remained. Influential astronomer William Huggins found that a nebula in Draco had an emission spectrum, characteristic of a cloud of thin, hot gas, rather than stars. He announced in 1864 that the "riddle of the nebulae was solved. The answer, which had come to us in the light itself, read: Not an aggregation of stars, but a luminous gas" (Harrison 1987, 115). He went further, proposing that the spiral nebulae are also clouds of gas and not made of stars. The pendulum thus swung back in favor of the Milky Way as the only large star system in the universe. By the turn of the twentieth century, many astronomers accepted that all nebulae were contained within the Milky

Way, and that the sun occupied a central or near-central position within the galaxy. In 1901, Dutch astronomer Jacobus Kapteyn used star counts, distances, and motions, to develop an updated disk model for the Milky Way (which he believed to be the only galaxy). The Milky Way was taken to be some 9,000 pc wide and 3,000 pc thick, with the sun located near the center. The influence of this model can be seen in Agnes Clerke's popular-level book, *The System of the Stars* (1905):

The question whether nebulae are external galaxies hardly any longer needs discussion. It has been answered by the progress of research. No competent thinker, with the whole of the available evidence before him, can now, it is safe to say, maintain any single nebula to be a star system of co-ordinate rank with the Milky Way. (Crowe 1994, 214)

Fortunately for the progress of astronomy, the self-correcting nature of science was about to turn this smug but erroneous opinion on its ear.

THE GREAT DEBATE

Despite the pronouncements of Agnes Clerke, not all astronomers had abandoned the island universe hypothesis. By the turn of the twentieth century, it had spawned two versions: in the first, the spiral nebulae were thought to be objects outside the Milky Way but much smaller than it (truly "islands"), while in the second, the spiral nebulae were considered Milky Ways of their own (or "comparable galaxies"). In his 1919 popular-level review of the status of cosmology, Hector Macpherson described it in this way:

In the present state of our knowledge we may compare our galactic system to a continent surrounded on all sides by the ocean of space, and the globular clusters to small islands lying at varying distances from its shores; while the spiral nebulae would appear to be either small islands, or else independent "continents" shining dimly out of Immensity. (1919, 334)

Not all the evidence used against the island universe hypothesis turned out to be good science. In 1916, Adriaan van Maanen of Mount Wilson Observatory announced that he had photographic evidence of rotation in M101, a spiral nebula seen face-on. Between 1921 and 1923 he published similar evidence for six other spirals. The problem was not that the spirals were presumed to rotate, but the size of the measured motion. If the spiral nebulae were outside the Milky Way, the observed rotations would correspond to unnatural velocities, some larger than the speed of light! This was taken by some astronomers as irrefutable evidence that the spiral nebulae could not be distant Milky Ways. However, as we shall see, these measurements would later be proven to be erroneous.

One astronomer greatly influenced by Van Maanen's supposed observations was Harlow Shapley of Mount Wilson Observatory (later Director of the Harvard College Observatory). His rejection of the island universe hypothesis was greatly influenced by his loyalty to his friend Van Maanen. Shapley had done his own calibration of the period-luminosity relation for Cepheid variables, and had used it, and other methods, to estimate the distances to the globular clusters visible in our galaxy. From this he found that they occupy a spherical halo surrounding the galaxy, centered around a point in the constellation of Sagittarius (where the visible gas and dust are noticeably thicker than average). Shapley therefore removed humans from a point of centrality within the galaxy, as the sun was now just another star orbiting at what Shapley estimated to be 20,000 pc from the center. In addition, his observations suggested a diameter for the Milky Way of about 60,000–90,000 pc, 10 times larger than previous estimates. If the Milky Way were such an immense object, and the spiral nebulae were other Milky Ways, it would mean that they, too, would be enormous, and the universe as a whole would be of inconceivable size. This seemed to argue against the island universe hypothesis.

On the other side of the argument was Heber Curtis of Lick Observatory (also in California). He and others had found novae in spiral nebulae, and in 1917 attempted to use them to estimate the distances to these objects. Using the fact that the average maximum apparent magnitude of novae in spiral nebulae was ten magnitudes dimmer than those in the Milky Way, he estimated that the spirals were approximately 100 times further away than the Milky Way's novae. A single, significant fly in the ointment was known to the astronomical community. In 1885 an unusually bright nova had been seen in the Andromeda Nebula. Its brightness suggested that either M31 was very close, or the object, dubbed S Andromedae (S And), was an unusual type of nova. A resolution was not achieved until 1920. Despite the problem posed by S And, Curtis was a vocal advocate for the island universe hypothesis. In 1917 he wrote "Our present evidence, so far as it goes, leads to the belief that the spirals are composed of great clouds of stars so infinitely distant that we cannot make out the individual stars." He also correctly explained the zone of avoidance (where no spiral nebulae could be seen in the sky) as being caused by "a great ring of absorbing matter somewhat like those which are found in certain edgewise spirals" (van den Bergh 1988, 9).

Curtis also realized that such absorbing material would dim the novae seen in the spiral nebulae, making them seem more distant than they would otherwise appear, but had no definite suggestion as to how to account for this effect. He described the island universe hypothesis in general as a "wonderful, a brain-staggering conception, more tremendous even than any other of the mighty ideas and facts of astronomy, that our own stellar universe may be but one of hundreds of thousands of similar universes" (van den Bergh 1988, 10).

In 1920 a public debate was arranged between Harlow Shapley and Heber Curtis on "The Scale of the Universe." The two issues under discussion were the size of the Milky Way and whether or not spiral nebulae were other Milky Ways. On April 26, 1920, Shapley and Curtis gave individual talks in the Baird Auditorium of the National Museum of Natural History in Washington, D.C., the event being later known as "The Great Debate." Afterward, both men submitted their revised comments for publication in the *Bulletin of the National Research Council*. The thrust of Curtis' argument was that the Milky Way was close in size to Kapteyn's estimates, and that star clusters and nebulae, with the exception of spiral nebulae, are contained within the galaxy. The spiral nebulae are Milky Ways in their own right, some 150,000 pc distant or more. Shapley held to his estimates of a much larger size for the Milky Way, and suggested that the spiral nebulae are much smaller than our galaxy. In addition, they might have been created within our galaxy but then flung outward some distance (which might explain the zone of avoidance). In future years, it was discovered that neither man was entirely correct. Although the spiral nebulae are indeed other galaxies, Shapley's estimate for the dimensions of the Milky Way was grossly oversized, and Curtis' too small.

THE DEVELOPMENT OF HUBBLE'S LAW

Two impediments to the acceptance of the island universe hypothesis were soon addressed by Knut Lundmark of Sweden. In 1920 he used 22 novae in M31 to estimate its distance to be about 200,000 pc. While this was too small by a factor of four, it was still much bigger than even Shapley's bloated estimate for the size of the Milky Way. This suggested that M31 was located a significant distance outside the Milky Way. As for the mysterious S And, Lundmark recognized that it represented a different type of object, what we today call a supernova, and there was no problem reconciling it with the distance calculated using the "normal" novae. Several years later, while visiting Mount Wilson, he utilized van Maanen's equipment and could find no evidence for rotation in the same spiral nebulae van Maanen had measured. To this day it is a mystery as to what exactly van Maanen thought he saw. In addition, Shapley's enlarged estimate for the size of the Milky Way was corrected in the 1930s to a more modern value, when the effects of interstellar dust were finally taken into account.

The use of novae in estimating distances to M31 prompted the young astronomer Edwin Hubble to do the same using Cepheid variables. Hubble had received his Ph.D. in astronomy in 1917 from the University of Chicago. His thesis had been a photographic study of spiral nebulae, and his conclusion was that they were separate galaxies from our own. In 1923 he discovered the first Cepheid variable in M31 using the 100-inch Hooker Telescope at Mount Wilson, and later found more, as well as Cepheids in spiral nebulae M33 and NGC 6822. NGC refers to J.L.E. Dreyer's *New General Catalog*, an 1887 revision of John Herschel's catalog. With its later supplement catalog, it listed over 9300 objects. Using Shapley's calibration of the period-luminosity relationship, Hubble was able to demonstrate that these objects were clearly outside the Milky Way, but was slow to publish his results, out of respect for Mount Wilson colleague van Maanen. Nevertheless, his results were presented, in his absence, on January 1, 1925, at a meeting of the American Astronomical Society. It should be noted that his distances were far too small (by about a factor of 2.5) due to the still-incorrect calibration of the period-luminosity relationship.

Once he had determined that the spiral nebulae were external to the Milky Way, Hubble became interested in their motions relative to each other. In 1912 Vesto Slipher of Lowell Observatory began measuring the Doppler shifts (and hence radial velocities) in the spectra of spiral nebulae and found that they were moving much faster than individual stars in our galaxy. He also found that the majority showed a redshift, which meant that they were receding from the Milky Way. The work was incredibly painstaking, and it took Slipher until 1925 to obtain results from 41 galactic spectra. During this time, Hubble was able to extend his distance measurements to farther galaxies by using the Cepheids he had observed in nearer galaxies. He was able to use the distances of these closer galaxies to estimate the absolute magnitudes of the brightest stars in these galaxies as well as the overall average absolute magnitudes of brighter galaxies. From these values he could determine the distances to galaxies out to the Virgo Cluster using the distance modulus formula. In a 1929 paper he plotted his distances versus Slipher's radial velocity data for several dozen galaxies and found "a linear correlation between distances and velocities" (Hubble 1929, 170). In mathematical language,

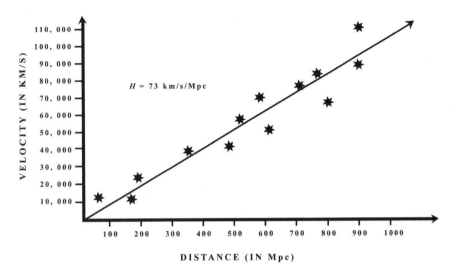

Figure 2.2 Hubble's Law

he found that velocity = distance × some constant, where, as we see in Figure 2.2, the constant is nothing more than the slope of the graph. This implied that the farther away a galaxy was, the faster it seemed to be receding from us. The result was surprising on several levels, but as Hubble (1929, 170) himself noted, "For such scanty material, so poorly distributed, the results are fairly definite." By 1935 Hubble and Milton Humason had extended the relationship to include a hundred more galaxies, out to distances far beyond the Virgo Cluster, and there was no doubt that the galaxies were receding away from each other.

The island universe debate seemed to be definitively closing. But one lingering issue remained, especially at Mount Wilson, namely van Maanen's supposed rotation of the spiral nebulae. Despite Lundmark's inability to reproduce van Maanen's observations, there were those loyal to their Mount Wilson colleague. Hubble's own patient respect for van Maanen finally reached an end:

They asked me to give him time; well, I gave him time, I gave him ten years. When a speaker at the R.A.S. [Royal Astronomical Society] announced if it were not for van Maanen's measurements, Hubble's results might be accepted, I decided to make the measurements. (Hetherington 1993, 357)

He remeasured the galaxies, searching for evidence of rotation and as he expected found nothing. This last impediment removed, Hubble had solved one significant problem in cosmology, but as is often the case in science, his discovery now raised several more questions. What was the physical cause of this motion? Did it suggest that the Milky Way was the center of the universe? Was there some other possible explanation for the redshift other than radial motion? The relationship eventually became known as *Hubble's law*, the motion of the galaxies it represents the *Hubble flow*, and the constant, the Hubble constant. In an interesting twist of vocabulary, as we shall see, the Hubble constant is really not constant, but changes as the universe evolves. What is calculated by astronomers is the value of Hubble's constant now (H_0), or at any other particular era in the universe's history. By measuring the change in the Hubble constant over time, much information can be gleaned about the past (and future) of the universe.

3

Warped Space-time: General Relativity and Modern Cosmology

GENERAL RELATIVITY

For two centuries, the Newtonian model of the cosmos—static, eternal, and ruled by gravity, a mysterious attractive force which derived from some property of mass—served as the basis for all cosmological research. Space and time were considered different beasts altogether, and there existed absolutes of both against which any measurements could be compared. With the publication of his general theory of relativity in 1915, Albert Einstein forever changed our understanding of the very nature of space and time, utterly shattering these absolutes in favor of an interconnected *space-time* in which observers can and do measure time and distance dependent on their positions in a warped background.

There is an old joke that only three people in the world understand general relativity, and one of them is Einstein himself. While general relativity is based on rather complex mathematics, its basic concepts and consequences can be understood with the help of analogies. As with any analogies, we must take care not to take them too literally, for if we do, we risk being led to absurdities. Although we experience space and time very differently, Einstein's theory treats them as four dimensions of something called space-time. This can be pictured as a flexible fabric made of three dimensions of space and one of time, interwoven in a very specific way called the *Einstein field equations*. The various ways that this fabric can be stretched and warped correspond to different mathematical solutions of this set of equations. Like any equations, there are two sides of the field equations that are set equal to each other. One side contains information on the energy and mass in that region of space-time. The other side of the equations contains information on the geometry of that space-time—how it is deformed by the presence of that mass and energy. Therefore these equations relate the shape of space-time to the amount and distribution of matter and energy present in the model. A helpful two-dimensional analogy is a rubber sheet. When a heavy object, such as a bowling ball, is placed into the center of the sheet, it bends or deforms in a particular way in response to the size, shape, and mass of the ball. Similarly, the presence of the sun deforms the "fabric" of the solar system, the resulting shape determining the orbits of the planets. In the often-quoted (and seemingly Zen) description by physicist J.A. Wheeler, "Matter tells space-time how to curve, and space-time tells matter how to move."

If your head is swimming, imagine how the scientific community of 1915 must have felt. What Einstein was proposing was nothing less than mind-boggling. Gravity was no longer to be pictured simply as a force like electricity, but as an entirely new type of phenomenon. Einstein understood the revolutionary nature of his theory, and as a good scientist, proposed three famous experimental tests which could be used to support or negate his claims: the precession of the perihelion of Mercury, the bending of starlight, and gravitational redshift.

Astronomy textbooks draw the orbits of planets as perfect ellipses, always returning to exactly the same point in space nearest the sun (or perihelion) at the end of one orbital period. In reality, the tugging of the planets on each other makes the orbits wobble, or precess. The resulting rosette-like pattern will be familiar to readers of the author's generation as the classic "spyrograph" pattern. Mercury, the closest planet to the sun, has the largest wobble of its perihelion. In 1859, Urbain Leverrier, who predicted the existence of Neptune due to its gravitational effects on the orbit of Uranus, found that after taking into account all the known planets, there was still a leftover precession of 43 seconds of arc per century in Mercury's orbit. Astronomers were at a loss to explain this, some going so far as to predict the existence of another planet, dubbed Vulcan, that supposedly orbited on the far side of Earth's orbit and remained hidden by the glare of the sun at all times. General relativity predicts that the warping of space-time by the sun should contribute precisely the observed amount of precession to Mercury's orbit.

According to Newtonian gravity, the path of a ray of starlight should be deflected by a tiny amount as it passes close by the sun. General relativity predicts that the amount of deflection should be twice the amount predicted by Newton. This would be detectable as a slight shift in the relative positions of stars appearing on the edge of the sun as compared to the same location in the sky when the sun was elsewhere. The practical problem is, of course, that one can't see stars appearing near the sun in the sky because the sun is too bright! The one exception is a total solar eclipse. British astronomer Arthur Eddington was an early supporter of general relativity and urged that the eclipse experiment be done. Astronomer Royal Sir Frank Dyson sent Eddington and colleagues to view the May 29, 1919, eclipse on the island of Principe, off the western coast of Africa, while another team led by Andrew Crommelin viewed the eclipse from Sobral in northern Brazil. The measurements of the two teams confirmed that starlight was bent by the amount Einstein had predicted. Their success is astounding if one considers that the images on the photographic plates were only deflected by a few hundredths of a millimeter! (Einstein 1961, 146–147). The announcement of the experimental results was reported in the *London Times* and other widely read newspapers, soon making Einstein a household name.

The third classical test also involves starlight, but in a different way. Since a star warps space-time, its light has to climb out of this "gravity well" in order to reach our eye. In doing so, the light loses energy, which means its wavelength becomes longer. This is called the gravitational redshift. For normal stars like the sun, the effect is miniscule, but for very dense stellar corpses, like white dwarfs and neutron stars, the effect is more important. It was first measured by Walter Adams in 1925, in the spectrum of the white dwarf companion of Sirius (jokingly nicknamed the Pup, as Sirius is known as the Dog Star). A black hole can be thought of as an object whose gravitational field is so intense that it creates an infinite gravitational redshift, as the light cannot escape at all!

Physicists and astronomers have since found numerous other tests of general relativity. These include the Lense–Thirring effect, in which the twisting of space-time by the earth's rotation affects the orbits of satellites in a specific way. This is currently in the process of being experimentally verified. In a textbook case of the scientific method in action, general relativity has withstood every experimental test thrown its way, and all of its major predictions have come to pass, with one lingering exception—*gravity waves*. In 1916 Einstein noted that distinctive ripples in the fabric of space-time should be created when masses move. Earthbound gravity wave detectors have so-far failed to directly detect gravity waves, and satellite detectors are still in the planning stages. In 1993 Joseph Taylor and his former graduate student, Russell Hulse, won the Nobel Prize in Physics for indirect evidence of gravity waves. Hulse had discovered a binary system comprised of two neutron stars, and over the years their orbit has become smaller in just the way predicted by general relativity. As the motion of the dead stars generates gravity waves, the pair loses orbital energy, resulting in a 3-ft per year shortening of their orbit.

COSMOLOGICAL MODELS

Modern cosmology is considered to have begun in 1917, when Einstein applied general relativity to the universe at large. He began his cosmological considerations with two basic assumptions, which he described as follows:

1. There exists an average density of matter in the whole of space which is everywhere the same and different from zero.
2. The magnitude ("radius") of space is independent of time.

(Einstein 1961, 152)

His first assumption, that matter and energy are distributed in an even manner, predicts that the universe looks the same at all locations and in all directions (it is homogenous and isotropic), which is referred to as the *cosmological principle*. This greatly simplified his field equations, but had no observational evidence to support it at that time. The second assumption, essentially that the universe was static, eternal, and unchanging, "appeared unavoidable" to Einstein (1961, 153), who "thought that one would get into bottomless speculation if one departed from it." To his surprise, Einstein found that his equations predicted that the universe was nonstatic, either expanding or contracting, so he added a "fudge factor" called the *cosmological constant*, which he admitted "was not required by the theory as such nor did it seem natural from a theoretical point of view," but instead was included to prop up the assumption of a static universe (Einstein 1961, 152).

Einstein's model universe was closed in on itself into a sphere, like the surface of the earth (but in a higher dimension), and was finite (limited) in size but unbounded (had no edge). It also fell prey to Olbers' paradox. Light rays would circulate around the universe, returning to their initial starting point, and over time such a universe filled with stars would become extremely bright.

In 1917 Dutch astronomer Willem de Sitter proposed another solution to the Einstein equations, a universe which also had a cosmological constant, but was devoid of any matter and yet was still curved. Einstein regarded the solution with suspicion, because he thought it was impossible to curve space-time without matter. He eventually came to accept the idea as possible, but because it was empty, probably not physically relevant. In fact, because of its emptiness, the solution might have been considered nothing more than a mathematical curiosity if it were not for one peculiar feature. If test particles of matter are sprinkled in a de Sitter space, they move apart at exponential speeds, producing a redshift that increases with distance. Because de Sitter originally considered his solution as static (because of the cosmological constant), this apparent motion seemed contradictory, and although de Sitter acknowledged this result, he did not investigate it sufficiently to be given credit for predicting the expansion of the universe discovered by Edwin Hubble. Hubble was certainly aware of this prediction of the de Sitter model, and in his 1929 paper, allowed for the speculation that "the velocity-distance relation may represent the de Sitter effect" (173).

In 1922, Russian mathematician Alexander Friedmann found a mathematical error in Einstein's work. In trying to prove that the universe must be static, Einstein had committed a well-known mathematical sin, namely dividing by a constant which could, in certain cases, be zero. Therefore the universe was not constrained to being constant. Friedmann published solutions which were nonstatic, or changing with time, including one which expanded from an initial singularity (like that found at the center of a black hole), and another which was periodic in time, expanding for a while then contracting. This "phoenix universe" was later studied in detail by Richard Tolman in the 1930s and became known as the oscillating universe. Although Einstein came to support this model, many questions remained as to whether it was physically possible.

With the publication of several papers by Friedmann, Einstein began to see the folly of his fudge factor. In a 1923 letter to mathematician Hermann Weyl he admitted that "if there is no quasi-static world, then away with the cosmological term" (Krauss 1999, 36). Einstein later called it "the biggest blunder of my life," and in a 1947 letter to cosmologist Georges Lemaître, admitted "Since I introduced this term I had always had a bad conscience. I am unable to believe that such an ugly thing is actually realized in nature" (Perivolaropoulos 2006, 16). Before we condemn Einstein for making such an error, recall the scientific method. Scientists routinely develop hypotheses which are then tested against observations and further theoretical models. Although one can argue that Einstein himself knew that the cosmological constant was an artificial construct from the first moment of its creation, at the time of his calculations (1916–1917) there was no solid observational evidence that the universe was other than static.

Friedmann had demonstrated that both Einstein and de Sitter's models were special cases of cosmologies allowed by general relativity, and had suggested wider classes of solutions that satisfied the cosmological principle. In 1935, H.P Robertson and A.G. Walker independently generalized Friedmann's work to create a model known as the

Friedmann–Robertson–Walker model. The central equation related the evolution of the universe in size (including, for example, the current expansion shown in Hubble's law) to the density of matter and energy in the universe, and the "curvature" of the universe (as well as any cosmological constant, in the most general case). In this model, whose analogies are represented in Figure 3.1, the universe has three overall (global) "curvatures" or geometries:

1. Euclidean or *Flat*—The universe can be pictured (in two dimensions) as the surface of a flat rubber sheet. As we learned in high school geometry class, two lines (including light beams) which begin parallel remain parallel no matter how far they travel. Euclid also demonstrated that the sum of the interior angles of a triangle equals 180 degrees. This geometry can be thought of as having zero curvature. This is the simplest cosmological model that actually has matter, and it is strangely referred to as the Einstein–de Sitter model, despite the fact that it differs from the models proposed by either cosmologist.

2. Spherical or *Closed*—This is similar to Einstein's original cosmology, in that the universe is pictured (in two dimensions) as the surface of a sphere. Lines or light beams which begin parallel will eventually merge (just as lines of longitude on the earth are parallel at the equator but meet at the poles). A triangle drawn on a sphere is distorted (bulges), such that the sum of the interior angles is greater than 180 degrees. This geometry has a positive curvature (as it curves in on itself).

3. Hyperbolical (Saddle) or *Open*—In this case the universe "flares out" like the center of a saddle, or a Pringles™ potato chip. Lines or light beams which begin parallel flare out as well, and a triangle drawn in the center of the saddle is "pinched," so the sum of the angles is less than 180 degrees. This geometry is said to have a negative curvature (the opposite of curving in on itself, or the opposite of a sphere).

These three geometries   correspond to three different fates for the universe, depending on the precarious balance between the attraction (pulling) of gravity and the repulsive (pushing) pressure of the hot early universe—what we today call the big bang. If the density of matter in the universe is below some critical value, nothing can counteract this initial push, and the universe will expand forever, although slowing its rate of expansion somewhat as time goes on. If the density of matter is above the same critical value, then the collective self-gravitation of the universe will counteract the expansion, and the universe will eventually

COSMIC GEOMETRY-CURVATURE
AND DENSITY

POSITIVE CURVATURE
HEAVY DENSITY

NEGATIVE CURVATURE
LIGHT DENSITY

ZERO CURVATURE
THE DENSITY IS JUST RIGHT

Figure 3.1 Three Possible Geometries of the Universe

collapse into what is called the big crunch. There is a possibility it could then reexpand, as in the case of Tolman's oscillating universe model. If the density of matter in the universe is precisely the critical value, then gravity and the big bang would be balanced, and the expansion of the universe would slow to a crawl over many billions of years, but never actually stop. This should remind the reader of the old story of Goldilocks and the Three Bears—one universe is too light (open), one is too heavy (closed) and one is just right (flat).

The special "just right" value for the density of matter in the universe is called the *critical density*, and cosmologists usually discuss it in terms of a ratio of the current density of matter (and energy) to this critical value. Therefore, in a flat universe the actual density is exactly equal to the critical density, so the ratio of the two (called Ω—omega, the last letter of the Greek alphabet) is precisely one. In a closed universe, the actual density is greater than the critical density, so Ω is greater than one, and in the open universe the actual density is less than the critical density,

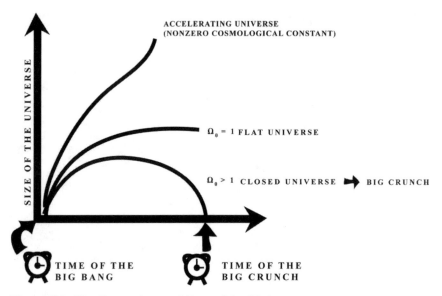

Figure 3.2 The Expansion and Fate of the Universe

so Ω is less than one. Therefore, if we could observationally determine the density of matter and energy in the universe, we would theoretically know its fate, assuming the value of the cosmological constant (if other than zero) is also understood.

This last caveat is because the relationship between geometry and destiny is more complex for the case where the cosmological constant is not zero. The original use of the cosmological constant to create a static universe depended on a perfect balance between the attraction of gravity and the repulsion created by the cosmological constant. But any hope at balance would only be possible (albeit not probable) if the cosmological constant is small and positive (greater than zero). As pictured in Figure 3.2, if it is large and positive it will not only counteract the tendency of gravity to pull the universe together, but the universe will expand forever and can actually accelerate as time increases, so long as the curvature is flat or negative. A spherical curvature and positive cosmological constant is Einstein's original model (without the artificial constraint that there be static balance), and depending on how things are fine-tuned, can expand forever or eventually recollapse. If the cosmological constant is negative in value (less than zero), the opposite occurs—the universe is doomed to collapse in on itself, regardless of the curvature.

We now return to several unresolved questions,   the first of which being the reason for the dark night sky. Recall that in 1861 von Mädler had argued that if the universe had a definite beginning in time, there would be a natural resolution for Olbers' paradox, because we would only receive the light from a finite number of stars. In a modern formulation of the problem, we must obviously replace "stars" with "galaxies," but the concept is identical. There has been some lingering confusion, however, with some modern textbooks giving equal credit for the dark sky to the finite age of the universe (and its galaxies) and the expansion of the universe. The argument is that since the universe is expanding, light is redshifted to wavelengths too long to see, thus making the sky darker than it should be. In an attempt to resolve the issue once and for all, in 1987, physicists Paul Wesson, K. Valle and R. Stabell did a detailed analysis of the contributions of both the finite lifetime of the universe and its expansion to solving Olbers' Paradox. Their results showed that the expansion only dims the light by a factor of about two, whereas the fact that the light from a finite number of galaxies has had time to reach us clearly plays the bulk of the role. In the words of the authors, "there can hardly be another astronomical subject that is so fundamental in nature yet so widely misunderstood" (Wesson, Valle, and Stabell 1987, 606).

This is also an appropriate time to clear up several common misconceptions about the expansion of the universe. Firstly, the galaxies are not receding away from us, but away from each other. We are not the center of the universe, nor does the concept of a geographic center have any place in modern cosmology. The universe is not expanding "into" something (such as more space). In the most literal sense, space itself is expanding, and as it does it carries the galaxies along for the ride. If this is the case, then what we have been calling the redshift really isn't the Doppler effect's redshift, but some effect that appears similar. This is certainly true, but the difference involves more mathematics than you really want to know, so we will pretend I haven't revealed that little secret to you. It is still true that the redshift is caused by the expansion of the universe, and measuring the redshift of different galaxies tells us how fast the universe is expanding at any given epoch during its life. It is also true that although objects within the universe cannot move faster than the speed of light, the fabric of the universe can stretch faster than light can travel. This does not violate any of Einstein's calculations, but means that there will be objects beyond our visible horizon that we will probably never be able to see. Some people consider these regions as other universes, since they cannot communicate with us. The daring reader is directed to the article by Max Tegmark listed in the bibliography for more information.

With all these possible cosmological models allowed by general relativity, is it possible to know with certainty which model describes the universe we currently inhabit? As part of that process of discovery, the serious implications of cosmological models that have singular beginnings led to the development of perhaps the most famous cosmological model of all—namely, the big bang.

THE BIG BANG AND STEADY-STATE THEORIES

Lemâitre and the Primeval Atom

Although Einstein began to regard the cosmological constant as nothing more than a scientific regret, other cosmologists continued to explore cosmological models which utilized it. One important example was published by Catholic priest and cosmologist Abbé Georges Lemâitre. Unfortunately, it first appeared in French, in a lesser-known journal, and only caught the attention of the scientific community 4 years later when an English translation appeared. In an interesting example of scientific provincialism, Lemâitre's paper both independently restated and extended Friedmann's model (of which he had been unaware) and provided the first observational evidence of what we now call Hubble's law, 2 years before Hubble's original paper on the subject.

Lemaître described the birth of the universe, from the radioactive decay of a large atomic nucleus, thusly:

We picture the primeval atom as filling space ... being nearly an isotope of a neutron. This atom is conceived as having existed for an instant only, in fact, it was unstable and, as soon as it came into being, it was broken into pieces which were again broken, in their turn.... An increase in volume resulted, the disintegration of the atom was thus accompanied by a rapid increase in the radius of space which the fragments of the primeval atom filled. (Munitz 1957, 343)

His model divided the history of the universe into three eras, the first being a rapid expansion due to the initial disintegration of his primeval atom. The second was a "coasting" epoch, in which the expansion rate dramatically slowed down, during which there was sufficient time for stars to form. During this phase gravity's attraction was more or less balanced by the cosmological constant. This was followed by,

the third period of accelerated expansion. It is doubtless in this third period that we find ourselves today, and the acceleration of space ... could well be responsible for the separation of the stars into extragalactic nebulae. (Kragh 1996, 52)

In this last stage, the cosmological constant dominated the evolution of the universe, driving the accelerating expansion. The astute reader may recognize this model as the first genesis of what later became known as the big bang theory—the idea that the universe had a defined beginning in a small, dense, hot state and has been expanding ever since.

Lemaître's model drew interest and criticism from the scientific as well as religious communities. Initial arguments against it included a disbelief that the laws of physics could possibly be reliably pushed back to the origin of the universe. Arthur Eddington found Lemaître's work important but troubling:

Philosophically the notion of a beginning of the present order of Nature is repugnant to me. I should like to find a genuine loophole. . . . As a scientist I simply do not believe the Universe began with a bang . . . it leaves me cold. (Singh 2004, 280)

Some religious figures, including Pope Pius XII, embraced it as scientific proof for the Judeo-Christian origin story. Lemaître himself struggled with the theological implications of his theory, but came down on the side of separate magisteria: "As far as I can see, such a theory remains entirely outside of any metaphysical or religious question" (Krauss 2005, 9). It is important to note that in terms of predicting the chemical make up of the current universe, Lemaître's model failed miserably, in that it suggested that most of the universe would be made of elements near the middle of the periodic table, such as iron. The problem was his use of the fission of a (very large) atomic nucleus as the initial state. Lemaître was on the right track, but it would take a better understanding of nuclear reactions in the early universe to derive a more realistic model.

Gamow's Evolutionary Model and the Steady-State

After World War II, Russian-born nuclear physicist George Gamow, alone and collaborating with Ralph Alpher and Robert Herman, investigated an improved model of the early universe in an attempt to discover the origin of the chemical elements. Rather than the decay of a primordial atomic nucleus, their model involved a hot, dense gas—dubbed ylem—that expanded and cooled. One particularly important paper was written by Alpher and Gamow, who added the name of colleague Hans Bethe to the byline to satisfy Gamow's well-known humor. The paper by Alpher, Bethe, and Gamow appeared in the April 1, 1948 edition of the normally humorless *Physical Review*, and became known as the $\alpha\beta\gamma$ paper. Gamow subsequently joked that he wished that Herman would change his last name to "Delter."

Several months after the publication of the $\alpha\beta\gamma$ paper, two papers by Cambridge scientists explaining a rival cosmology appeared in the same volume of the *Monthly Notices of the Royal Astronomical Society*. The first, by Hermann Bondi and Thomas Gold, laid out what they called the *steady-state theory*, in which the universe obeyed the *perfect cosmological principle*. This stated that the universe not only looks the same everywhere and in all directions, but the same at each instant of time. In other words, it is eternal and largely unchanging, although individual stars can, of course, be born and die. However, since Hubble's law meant that galaxies are moving farther apart from each other, the density of matter in the universe would naturally decrease with time. In order to combat this, new material would need to be spontaneously created in order to obey the perfect cosmological principle. This aspect of the steady-state model was explored in Fred Hoyle's paper. This would clearly violate the long-cherished concept of conservation of matter, but would occur at such a slow rate of speed (imagine creating a single hydrogen atom in a one liter soda bottle every trillion years) that it would be very difficult to see directly.

The stage was now set for another fundamental debate in cosmology, between what Gamow called the evolutionary theory and the steady-state theory. The steady-state appealed to those who, like Bondi and Gold, doubted that the laws of physics seen in the universe today could really be expected to hold under the extreme conditions demanded in the hot, dense state of Gamow's ylem, or in general wished to sidestep knotty questions about the origin of the universe. There was also an antireligious backlash against the evolutionary cosmological model, which reportedly was part of the motivation for the rival steady-state. The name by which Gamow's model eventually became known, the *big bang*, was actually first spoken by Fred Hoyle during one of his popular astronomy radio shows, and was meant as a serious put-down. On the other side, Gamow and Lemaître dismissed the steady-state out of hand, with Gamow referring to it as "artificial and unreal" (Kragh 1996, 297).

The Rise of the Big Bang

As with any scientific debate, the final outcome depends not on a popularity contest or democratic election, but on the scientific method. Observations would ultimately determine which candidate theory deserved to be taken seriously by the cosmology community. Hubble's law was a natural outcome of the expanding universe of the big bang, but could be accommodated, albeit only with the inclusion of the spontaneous creation of matter, in the steady-state. In fact, Hoyle and certain colleagues spent many years unsuccessfully trying to find an alternative explanation for the redshift of the galaxies. The latter model did put a serious constraint on Hubble's law. Since the universe had to obey the perfect cosmological principle and could not evolve on large scales, the value of Hubble's constant could not change over time. Conversely, the big bang model suggested that speed of the expansion should change over time, the exact rate of change depending on the over-all geometry of the universe. When both models were developed, the uncertainty in the graph of Hubble's law could safely incorporate either possibility. But by the late 1950s the data suggested that Hubble's law did not match the predictions of the steady-state model.

Another prediction of the perfect cosmological principle and the spontaneous creation of matter was that, just as stars of different ages could be found everywhere in the universe, individual galaxies had to be created at different points in the universe's history. However, all the galaxies seen within our galactic neighborhood appeared to have similar ages. Martin Ryle, one of the pioneers of radio astronomy, led the radio astronomy group at Cambridge, and like Bondi, Hoyle, and Gold had worked on radar during World War II. By 1962 he had found another violation of the perfect cosmological principle. Extragalactic radio sources, now known to be galaxies with voracious black holes in their centers, seemed more common the farther out in space his instruments probed, meaning they were more common the farther back in time one looked. Martin Ryle and others continued to find that the distribution of these radio galaxies and their relatives, quasars, did not match that predicted by a steady-state universe.

Since Gamow's motivation for the big bang theory was the search for an explanation for the formation of the chemical elements, the predictions of both theories should be checked against the observed composition of the universe. The big bang largely failed in its grandiose mission to build up all chemical elements by adding neutrons to successively heavier atoms, because there are no stable atomic nuclei with an atomic mass of five or eight. This essentially stops the sequence at helium. Despite this serious limitation, the big bang was exceedingly successful in predicting that the universe would have an overall composition of 75 percent hydrogen to 25 percent helium (by mass). The steady-state model had hydrogen being created through the spontaneous creation of matter, with all other elements being cooked inside stars. Demonstrating that this was possible was one of the main reasons behind the famous B^2FH paper (where the H stands for coauthor Fred Hoyle of steady-state infamy). The problem was that stellar nucleosynthesis alone could not account for the quarter of the universe that was helium. In a rather strange twist of fate, the big bang theory, when coupled with the B[2]FH paper which had been motivated by the steady-state theory, was very successful in explaining the observed chemistry of the universe and the origin of the chemical elements.

The strongest evidence for the big bang (and against the steady-state) was the theory's originally least-discussed prediction. Gamow predicted that as the universe cooled, there would come an important transition at which radiation (light) would no longer strongly interact with matter, but instead would be free to move about the universe. Alpher and Herman calculated that there would be a resulting background blackbody radiation currently at a temperature of around 5 K. The prediction was largely forgotten, possibly because such a weak signal was thought nearly impossible to detect. Intrigued by the oscillating cosmological model, which would have a series of big bangs and big crunches, Princeton physicist Robert Dicke set out in 1964 to detect this feeble radiation, which, according to Wien's law, would have its peak in the microwave section of the electromagnetic spectrum. To Dicke's chagrin, before he could assemble his detector, Arno Penzias and Robert Wilson, two engineers working for Bell Laboratories—like Princeton, in New Jersey—had accidentally discovered this "echo" of the big bang (a discovery which would earn them the 1978 Nobel Prize in physics), at a temperature of approximately 3 K. Companion papers by both Dicke and his team and Penzias and Wilson were published in May 1965, but to the enduring annoyance of Gamow, neither mentioned that this *cosmic microwave background* had previously been predicted. What is not widely known, even among astronomers, is that the cosmic microwave background narrowly missed discovery two decades before! In 1941, Australian physicist Andrew McKellar found that a peculiar spectral line in the star ζ Ophiuchi was due to the excitation of molecules of cyanogen in the interstellar space between the star and Earth, which had a temperature of around 2 K.

By the summer of 1965, it was becoming nearly impossible to support the steady-state model, and the big bang became the reigning model of the evolution of the universe. At a 1982 meeting of the International Astronomical Union, Russian cosmologist Yakov Zeldovich proclaimed that the big bang was "as certain as that the Earth goes round the Sun" (Rees 2000, 15). This has led to the scientifically erroneous claim that the big bang theory has been "proven." There is also the equally wrong idea that the big bang theory explains how the universe "exploded" from an initial singularity. It is more properly said that the expansion of the universe from a hot, dense state of being as explained by the big bang theory, especially its inflationary revision in the 1980s, is the scientific "best fit" to the current accumulated observations of the universe. How this initial state itself came into being, and whether it was ever truly a singularity, is the regime of further speculation and other theories.

4

MEASURING AND
MAPPING THE UNIVERSE

NORMAL GALAXIES

We now return to the considerable contributions of one Edwin Hubble.
As part of his study of galaxies and their properties, he published a
classification scheme in 1926 which divided them into four classes:
ellipticals, spirals, barred spirals, and *irregulars.* Each class was again subdivided, creating a continuum of galaxy classes (minus the irregulars)
sometimes represented in a horizontal Y-shape called the *Hubble tuning
fork diagram* as represented in Figure 4.1. Galaxies seen face-on are far
easier to classify than those seen edge-on, and some ambiguities (and
arguments) are inevitable.

When most people think of a galaxy, spirals like the Milky Way come to
mind. This is certainly understandable, as we have seen that the "spiral
nebulae" were the first objects that astronomers considered as possibly
existing beyond the Milky Way. Spirals are defined by two or more arms
that emanate from either a central nucleus or an elongated structure
called a bar. This led Hubble to differentiate between *normal spirals*
(sometimes just called spirals) and *barred spirals.* The ratio of barred
to "unbarred" is highly debated by astronomers, with estimates varying
from 30 to 60 percent of all spirals having central bars. Both types
of spirals were subdivided by Hubble according to three criteria: the
prominence of the nucleus (also called the central or nuclear bulge),
the tightness of the winding of the spiral arms, and the clumpiness of
the material within the spiral arms. Sa and SBa (SB denoting spiral-
barred) galaxies have prominent nuclear bulges, and tightly wound
arms that appear to have a smooth distribution of stars. Sb/SBb galaxies
have intermediate properties, and Sc/SBc galaxies have "inconspicuous
bulges with open and fragmented spiral arms that are highly patchy"

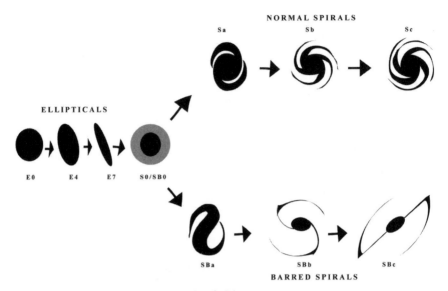

Figure 4.1 The Hubble Tuning Fork Diagram

(Waller and Hodge 2003, 11). There are also Sd galaxies (not part of Hubble's original scheme) in which the nuclear bulge is unrecognizable and the spiral arms hard to distinguish.

Several different populations of stars inhabit spiral galaxies, from very old extreme population II stars in globular clusters to newborn population I stars being created from the copious gas and dust in the spiral arms. In "early type" barred spirals (SBa), the bars are composed of stars, while in "late type" barred spirals (SBb and SBc) there is a significant proportion of gas and dust in the bar as well. Bar-like structures are sometimes seen in irregular galaxies (such as the Large Magellanic Cloud), which has led some authors to acknowledge another type of spiral, the Magellanic type (Sm and SBm), which can have one arm. Other sources classify these galaxies under the irregular category.

At the joint in the Hubble tuning fork diagram, where the two types of spirals meet, we find the *lenticular* ("lens-shaped") *galaxies.* They resemble spirals in that they have a central bulge (and sometimes a central bar as well), as well as a truncated disk emanating from the bulge, but they lack the spiral arms that dominate the disks of spiral galaxies. Lenticulars without bars are called S0, while those with bars are termed SB0.

Passing through the lenticulars to the last arm of the tuning fork, we find the elliptical galaxies. Hubble subdivided them according to how elliptical they appeared, with E0 galaxies being spherical and E7 highly elongated (sometimes referred to as "cigar-shaped"). They are rather featureless and tend to have less gas and dust than spirals, resulting in few young stars or star-forming regions. Elliptical galaxies show the greatest variety in size of any class of galaxies, from the humongous cD ellipticals down to the puny dwarf ellipticals and dwarf spheroidals. The centers of rich clusters of galaxies, where the galaxies are more densely packed, are typically dominated by a huge cD elliptical. The name is shorthand for supergiant-diffuse, which accurately describes their appearance. These monsters can be dozens of times larger than our Milky Way and are believed to be built up from numerous mergers of smaller galaxies. Some contain evidence of multiple nuclei, the remnants of previous galactic meals.

On the other end of the size scale exist dwarf galaxies, some no brighter than a globular cluster. Dwarf ellipticals and their smaller and dimmer cousins, the dwarf spheroidals, appear to be the most common type of galaxies, making up at least 60 percent of all galaxies. Their contribution might be as high as 85 percent, but it is easy to overlook small and dim objects when doing surveys of the universe. Dwarf ellipticals differ from the globular clusters they somewhat resemble in that they contain a unique type of variable termed anomalous Cepheids. They are approximately a billion years old, far younger than stars seen in globular clusters. Dwarf spheroidals are several hundred parsecs across, far larger than any globular cluster, but have a similar overall luminosity. Spreading the light over a much larger area makes these galaxies exceedingly diffuse and difficult to see. It is therefore no surprise that dwarf spheroidal companions to the Milky Way continue to be discovered.

Irregular galaxies generally lack spiral or elliptical structure, as their name implies. They tend to be smaller than spirals, and also come in dwarf varieties. They contain gas and dust in sufficient quantities to have young stars and regions of ongoing star formation. They are sometimes found close to other galaxies, leading to the speculation that their disturbed shapes might be due to gravitational interactions among galaxies.

Given that galaxies come in a variety of shapes and sizes, scientists have searched for a connection between Hubble type and a galaxy's birth and evolution. The actual process(es) and timing(s) of galaxy formation in our universe are areas of active research, but as strange as it may seem, in some ways astronomers know less about the evolution of galaxies than they do the evolution of stars or the universe at large. Fortunately, astronomers can use look-back time to their advantage. By taking long-exposure photographs, peering as deeply into space as possible, galaxies can be studied at younger and younger times in their lives. For example, the Hubble Ultra Deep Field (HUDF) photograph captured the images of some 10,000 galaxies. It found that galaxies with redshifts larger than four, corresponding to the first few billion years of the universe, look completely different from the Hubble classes seen in the universe today.

Clues as to how the various Hubble classes might have arisen can be found by comparing spirals and ellipticals seen today. Ellipticals currently have very few young stars and star forming regions, suggesting that the bulk of their stars were formed early on in the galaxy's history. By comparison, the arms of spirals contain active star forming regions, with Sc/SBc subclasses having the most vigorous current star formation. Galaxies with unusually high amounts of ongoing star formation are called starburst galaxies, and are discovered by the enormous amounts of infrared radiation this star formation spawns. In the 1980s, the IRAS satellite discovered that starburst galaxies emit 100 times more energy in the infrared alone than our galaxy does over all wavelengths. The trigger for this abnormally high rate of star formation is possibly the merger or gravitational interactions of two galaxies. If elliptical galaxies formed through mergers early in their history, it would have triggered such tremendous star formation that there would now be insufficient free gas and dust for future stars.

The stars in the disk and spiral arms of spiral galaxies have regular, orderly orbits around the center of the galaxy, assumed to be a relic from the formation of the galaxy. Elliptical galaxies lack this orderly rotation, and have been compared to "'star piles'—swarms of stars that have long since settled down into galactic orbits under one another's gravitational influence" (Jones and Lambourne 2004, 78). Computer models have shown that the rate of initial rotation of the cloud presumed to form the galaxy is key to whether a spiral or elliptical forms. If the cloud spins quickly, most of the material will tend to form a disk before a large number of stars form, leading to a spiral. If the cloud spins slowly, many stars will begin to form while the cloud has a roughly spherical shape and a discernable disk will not emerge, creating an elliptical galaxy.

Understanding how the flattened disk of a spiral galaxy forms still leaves us with one important question—how do the spiral arms not only form out of the disk, but remain intact billions of years after the galaxy's formation? Since spiral galaxies are not solid objects and the inner regions can move faster than the outer portion (called differential rotation), the spiral arms should wind up and dissipate over time. However, this is not observed to occur. The best explanation is the *density wave theory* developed in the 1960s by C.C. Lin and his students at MIT. The model is based on the premise that areas of slightly higher than normal density will form in the disk of the galaxy, creating "grooves" in the space-time fabric of the galaxy. As stars and gas clouds orbit the center of the galaxy, they move into these gravitational troughs, slow down, and linger there for some length of time before moving on. The best analogy is a traffic jam, created because of rubbernecking near an accident scene. Individual cars slow down and clump together in the familiar "bumper to bumper" pattern defining the jam, but soon move onward. The traffic jam is fixed in space, but individual cars move in and out of it. Likewise, individual stars and nebulae move in and out of the spiral arms. While in the "jam" of the spiral arms, nebulae can be compressed, triggering star formation. This explains why the spiral arms are areas of ongoing star formation.

Computer models have not only shown that bar-like structures in galaxies are relatively easy to make, but are in fact so easy that it is much harder to explain spirals that lack them. Like spiral arms, it is believed that individual stars move in and out of the bar, which itself stays fixed in its relative position within the galaxy. In keeping with these concepts, the next topic to explore is the structure of the Milky Way itself and its neighborhood—do we really live in a (normal) spiral galaxy?

THE MILKY WAY

As seen in our historical survey, it has been a serious challenge ·to map the overall structure of the galaxy from our fixed position within it, made even more difficult by the gas and dust blocking our view in some directions (the zone of avoidance). According to current observations, our home galaxy is composed of about 10^{11} stars, most of which inhabit the flattened disk. This disk has a radius of at least 15 kpc (15,000 pc) and we are located about 8,500 pc from the galactic center. The stars of the disk are divided into the thin disk component (closer to the "equatorial plane" of the galaxy) and thick disk component (located just above and below the thin disk). The majority of the gas and dust is found within the thin disk. The disk has a slight warp near its edge, a property found in roughly a quarter of spiral galaxies.

The sun orbits the center of the galaxy with a period of about 250 million years. Its motion is rather complex, tracing out a "rosette-like" pattern. For example, the sun moves slightly closer and farther away from the center with an estimated period of 169 million years, and bobs up and down through the plane of the galaxy with a period of about 62 million years (Waller and Hodge 2003, 110). Using regions associated with young stars and star formation, such as OB associations and HII regions, astronomers have traced out three spiral arms in our neighborhood: the Sagittarius–Carina arm (looking toward the center of the galaxy), the Orion–Cygnus arm, and the Perseus arm. The sun is located on the edge of the Orion–Cygnus arm, a structure that is small enough to sometimes be considered merely a spur of a larger arm.

As many as 95 percent of the disk stars are found within the thin disk, as depicted in Figure 4.2, a region extending a few hundred parsecs above and below the central plane of the galaxy. Stars specifically residing within the spiral arms are the youngest stars in the galaxy (population I), with ages of 100 million years or less. The rest of the thin disk population is older, with ages between 1 and 10 billion years (including the sun). The older stars of this group are found at higher galactic latitudes (distances above and below the plane of the galaxy), suggesting that stars tend to be formed closer to the plane and later migrate outward, perhaps due to gravitational interactions. Within a few thousand parsecs of the galactic plane are found the older stars (around 10 billion years) of the thick disk, and are called intermediate population I. It has been suggested that these stars also formed closer to the central plane, but their orbits were moved to higher latitudes by gravitational interactions as well. In other words, over time the disk of the galaxy "puffs out," forming several populations.

Surrounding the galaxy's disk is a halo, in the shape of a slightly flattened sphere. The exact size of the halo is difficult to measure, but it extends out well past the disk. The most obvious citizens of the halo are over 150 globular clusters, the majority of which are made of very old, metal-poor stars (extreme population II). Isolated halo stars are also found, including RR Lyrae stars, and are also old and metal poor. The spherical shape of the halo and the swarming motion of its stars in highly elliptical orbits well out of the plane of the galaxy, suggest that the halo was the first portion of the Milky Way to form, as the rapidly rotating protogalactic cloud first began collapsing into the disk.

Looking toward the center of the galaxy (in the general direction of the constellation Sagittarius), the nuclear bulge comes into view. The stars found here are old, nearly as old as those in the halo, but are chemically distinct. Although they are normally classified as population II stars, they have metallicities much higher than halo stars, and in some cases have the same chemistry as the sun. The stars and HII regions in the bulge have more highly elliptical orbits than the stars of the disk, which led Gerard deVaucouleurs to suggest in 1964 that the Milky Way might be a barred spiral. Further evidence has accumulated through infrared studies, which are able to pierce through the intervening gas

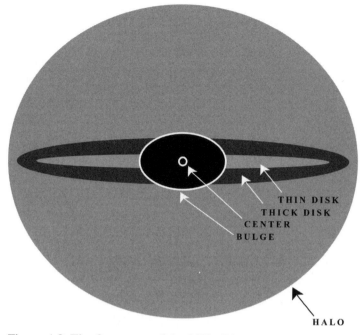

THIN DISK
THICK DISK
CENTER
BULGE

HALO

Figure 4.2 The Structure of the Milky Way

and dust. The Galactic Legacy Infrared Mid-Plane Survey Extraordinaire (GLIMPSE) found about 30 million IR sources between the sun and the center of the galaxy. Based on their distribution, it appears that the Milky Way has a central bar about 9 kpc long and a few kpc wide, tilted by 44 degrees to the direct line between the sun and the galactic center. Now that we have ascertained that the Milky Way is probably a barred spiral, what about its Hubble subtype? Based on the size of the nuclear bulge and the structure of the spiral arms, a Hubble class of SBbc (between an SBb and SBc) is suggested.

The intervening gas and dust between our solar system and the center of the galaxy screen it from view, at least at visible wavelengths. Whatever lies hidden at the center of the galaxy generates a tremendous amount of energy at infrared, x-ray, radio, and gamma ray wavelengths. The exact center of the galaxy is believed to be marked by a strong radio source named Sgr A*. Motions of stars near the center suggest that there is a very compact mass of about two to three million solar masses contained within an area of 0.001 pc. The only sensible astronomical explanation appears to be a *supermassive black hole* (SMBH), with the energy being generated by its messy eating habits. Such SMBH appear to be common at the centers of galaxies, and the details of their energy generation will be discussed in an upcoming section.

The structure of the galaxy, including the chemistry of its various sections, should be explained by any model of galactic formation. Since the halo contains the oldest stars, it makes sense that it was formed first. As the protogalaxy began to collapse from a relatively spherical shape, the first stars formed. The bulk of the material continued to collapse to form the disk, and star formation will occur later here, resulting in younger stars. But what of the nuclear bulge? It contains stars nearly as old as the halo, but with richer chemistry. Near the center of the original protogalaxy, material collapsed and aggregated to form the bulge. Star formation occurred very quickly here because of the high density, and the first (massive) stars died relatively soon, rapidly enriching the material incorporated in the next generations, leading to higher metallicities.

THE LOCAL GROUP

Just as stars are typically found in groups and clusters, many galaxies aggregate into a larger galactic neighborhood called a galaxy cluster. Clusters of galaxies with fewer than 50 members are generally called groups, as in the case of the Milky Way. Its extended family includes some several dozen members, and was termed the Local Group by Edwin Hubble. These 40 or so galaxies are contained within a diameter of 1.5 Mpc (million pc or megaparsec), and the intervening space between it and the nearest group—some three to four Mpc—contains very few galaxies. There are three large galaxies, all spirals: M31, the Milky Way, and M33. Most of the remaining galaxies are dwarf irregulars, ellipticals, and spheroidals. Most of these in turn cluster around either M31/M33 (which are relatively close together) or the Milky Way.

M31, the Andromeda Galaxy, is the largest and most luminous member of the Local Group, and is considered an Sb galaxy. It is the most distant object visible to the naked eye, and was first recorded by al-Sufi in the tenth century. It has at least 26 known satellite galaxies, as well as five objects which appear as extended globular clusters but in reality are possibly the remains of dwarf galaxies that were gravitationally cannibalized by their much larger family member. There is evidence of a secondary core in M31, which could be the result of the past merger with another galaxy. Our own Milky Way may merge with M31 in about 6 billion years.

The most massive satellite of M31 is M32, a small (but not dwarf) elliptical. Although it has less than 1 percent of the mass of M31, it is close enough for the two to interact gravitationally. The spiral arms on M32's side of the Andromeda galaxy are irregular, and it is thought that M32 was once larger, many of its stars having been ripped away by its bigger brother. M33 is an Sc spiral and is about one-third the diameter of M31.

Like M31, the Milky Way has a number of galactic companions (at least 14, with new candidates continually being discovered) along with several possible "stripped core" objects which are sometimes classified as globular clusters. The zone of avoidance has made it difficult to get a full census of the Milky Way's satellites, and low luminosity dwarf spheroidal companions continue to be discovered. For example, SagDEG (Sagittarius dwarf elliptical galaxy) was only discovered in 1994, hiding within the zone of avoidance. A mere 15 kpc from the center of the galaxy, it is doomed to be cannibalized by the Milky Way one day. Already falling victim to the Milky Way's gravity is Sag DIG (Sagittarius dwarf irregular galaxy), which lies on the far side of the galaxy. As it spills its stellar guts across the sky in a 10 kpc stream, it adds its globular clusters to our halo, their foreign identity revealed by their anomalous motions. An increasing number of tidal streams, composed of material stripped from satellite galaxies, have been found around the Milky Way and M31, demonstrating the importance of gravitational interactions between galaxies within clusters.

By far the most obvious members of the Local Group (outside of the Milky Way) are the Large and Small Magellanic Clouds (LMC and SMC). The LMC is less than half the size of the disk of the Milky Way, and the SMC is half as large as the LMC. The LMC is alternately called an irregular with a bar, or the prototype of the barred Magellanic spirals. During its first few billion years, it enjoyed an active period of star formation, including around a dozen globular clusters and stars that are now RR Lyrae variables. After a 7-billion-year hiatus, star formation was violently renewed, and continues to this day. A stunning example is the HII region called the Tarantula Nebula. The SMC is an irregular dwarf galaxy, with more gas than its larger sibling and only one globular cluster. These facts suggest that it is a "more primitive and less evolved" galaxy, and star formation may have "started off later, or more gradually, than it did in the Large Cloud" (van den Bergh 2000, 142).

The Magellanic Clouds interact gravitationally with each other, with the larger cloud doing more damage to the smaller one. Both of these interact with the Milky Way as well. Both Clouds have highly ellipti-cal orbits around the Milky Way with 2 billion year periods. Although they last passed close to the Milky Way hundreds of millions of years ago, roughly 50 million years ago they were only 10 kpc apart (half the distance they now keep from each other), and tidal disruptions were probable. There is a bridge of gas peppered with young star clusters connecting the clouds to each other, as well as an extended trail of nebulae spread across nearly half of the SMC's orbit, called the Magellanic Stream. Eventually both the Clouds will be devoured by the Milky Way, as their orbits slowly spiral in toward our galaxy. It appears that violent eating habits are the norm at the galactic level, and nowhere is this more apparent than the cores of distant galaxies, dominated by their voracious black holes.

ACTIVE GALACTIC NUCLEI

In 1943, Carl Seyfert noted that some spiral galaxies have unusually bright, compact cores. Upon studying their spectra, he also found that unlike normal galaxies, these galaxies had strong emission lines. These *Seyfert galaxies* were later found to come in two varieties, Type 1, in which the emission lines are broad, and Type 2, in which the emission lines are narrower. The difference in the width of the spectral lines is caused by *Doppler broadening*. Recall that when a light source approaches, its light is blueshifted, and when a light source recedes its light is redshifted. If a variety of light sources (such as stars or clouds of hot gas) are moving in various directions, some of the light will be blueshifted and some will be redshifted. If all the objects are emitting light due to the same element, then the spectral line will appear widened as some of the light is shifted one way and some the other. The wider (broader) the line, the faster the stars or gas clouds are moving. The broad lines of Type 1 Seyferts are caused by material moving faster than 1,000 km/sec.

Seyfert galaxies are also characterized by significant and sometimes swift variations in the amount of visible light and x-rays they emit. The short time scale of these changes—doubling in brightness in only a few hours—means that whatever engine is emitting the x-rays is only the size that light can travel in a few hours (or similar to the size of the solar system out through Neptune). A census of spirals suggests that 10 percent of Sa/SbA and Sb/SBb spirals are Seyferts. This can be interpreted in two ways: either only 10 percent of early-type spirals ever exhibit Seyfert-like activity, or all such spirals can go through a Seyfert phase for 10 percent of their current lifetime.

Elliptical galaxies can also exhibit strange behavior. In 1937, engineer and amateur radio astronomer Grote Reber built a radio telescope in his backyard and began mapping radio emissions from the Milky Way. In 1944 he reported the discovery of radio sources in Cassiopeia and

Cygnus, later named Cassiopeia A and Cygnus A. The first object was found to be a supernova remnant, while the second was associated with the center of an obscure galaxy, peculiar at visible wavelengths only in that it was crossed by dark dust lanes. Cygnus A is a classic example of a lobe-dominated radio source, where the radio emissions come from two gargantuan regions which extend several hundred kiloparsecs into space. Some *radio galaxies* show radio emission in more concentrated jets that seem to connect the vast distended lobes to the core of the galaxy. Radio galaxies with most of the emission coming from the jets are called Fanaroff–Riley Type I (FR I) while those dominated by the radio lobes are dubbed Fanaroff–Riley Type II (FR II). As in the case of pulsars, the radio waves are synchrotron radiation generated by electrons spiraling in powerful magnetic fields.

In 1960, astronomers set to work finding the visible counterparts to the radio signals listed in the Third Cambridge radio catalog. It was discovered that some sources, such as 3C 273 and 3C 48, did not seem to coincide with a supernova remnant or galaxy. 3C 48 appeared to be associated with a faint blue star, but stars normally do not emit significant amounts of radio waves. The star's spectrum was more puzzling still, as it had very obvious and broad emission lines (also not characteristic of stars) that did not correspond to any known elements. 3C 273 was found to be a similar object 2 years later. Astronomer Maarten Schmidt noticed that the unidentifiable spectral lines had a familiar pattern, that of the visible wavelengths of hydrogen, but were redshifted by such a large amount that no one had realized their identity. From Hubble's law, if these objects have very large redshifts, they must be located very far away, and the fact that they could be seen as star-like objects must mean they have high intrinsic luminosities. The term quasi-stellar radio source, or quasar, was quickly coined to describe them. The more generic *quasi-stellar object (QSO)* became used because a large number of similar objects do not have significant radio emissions, but the term quasar is still commonly used in popular level texts for all related objects.

For many QSOs, a faint host galaxy has been spotted, but due to their huge distances, it is extremely difficult to see these host galaxies clearly enough to determine their Hubble classes. All QSOs emit an unusually large amount of energy at infrared, ultraviolet, and x-ray wavelengths, as well as have high visual luminosities. Although QSOs were first identified through their radio emissions, about 90 percent are "radio quiet." These radio-quiet QSOs are found in both spiral and elliptical hosts. The broadening of the emission lines showed that material within the QSO is moving at up to 10,000 km/sec. In this and other ways, the spectra look similar to Type 1 Seyferts, leading to the comparison of radio-quiet quasars with "Seyferts on steroids" (Wiita 2006, 3). The fact that radio-loud QSOs are generally associated with elliptical and interacting galaxies leads to a similar conclusion in their relation to radio galaxies.

The extreme redshifts of QSOs and their corresponding vast distances leads us to the conclusion that we are looking at objects as they were long ago. Most have redshifts of two to three, corresponding to an era several billion years after the origin of the universe, and QSOs appear to have become more rare as the universe evolved. The most distant QSOs have confirmed redshifts of just under seven, corresponding to an era only 800 million years into the life of the universe. Whatever engine powers the QSOs must generate less energy with time, and possibly shuts off at some late date. Another aspect of QSOs with cosmological importance is how their light interacts with intergalactic space. Intervening clouds of cool hydrogen will absorb the Lyman $\acute{a}$ line as electrons jump from the ground state to the first excited state. But each of these clouds is at a different distance, both from each other and from the QSO itself, resulting in a series of closely spaced lines at slightly different redshifts called the Lyman $\acute{a}$ forest. In studying the spectra of distant QSOs, we get free information about material spread throughout the universe.

In 1926, German variable star astronomer Cuno Hoffmeister discovered a nondescript variable star which was given the ordinary name BL Lacertae (BL Lac for short). In 1968 it was discovered to be a strong radio source, and reclassified as a QSO. It was afterward reclassified once more, as the prototype of a new class of objects, the *BL Lac objects*, which are most often found associated with the cores of distant elliptical galaxies. Although they appear stellar like QSOs, these objects lack the strong emission lines found in QSOs and are generally strong and rapidly variable emitters of γ rays, the most energetic form of light. Their emissions in the visible and radio sections of the electromagnetic spectrum are also highly polarized. Some of these properties are shared by a subclass of QSOs called optically violent variables (OVV), and they and BL Lac objects are sometimes grouped together under the term *blazar*.

By the 1970s it appeared that there were a number of strange members of the cosmological zoo who shared certain similarities, including extreme energy output over many segments of the electromagnetic spectrum originating from a rather small area, large distance, and an association with the cores of galaxies. Could it be that the relationship was more than skin-deep? In 1985, R.R.J. Antonucci and J.S. Miller demonstrated that Type 2 Seyferts are nothing more than Type 1 Seyferts viewed sideways through a donut, or torus, of dust. This was the beginning of the unified model of *active galactic nuclei* (AGN) as pictured in Figure 4.3. At the heart of an AGN is a supermassive black hole millions of times the mass of the sun, similar to that found in the core of the Milky Way. This is the surprising engine that drives the energy source, despite the fact that nothing can escape from a black hole (at least above the quantum level). The key is the black hole's messy eating habits.

Recall that as material is drawn further into the intense gravitational field of the black hole, it forms a swirling mass called an accretion disk. Gas and dust particles move so fast and collide with such energy that the temperature rises to 1 million Kelvin, and x-rays and ultraviolet rays are emitted. These high energy wavelengths will heat any clouds of gas orbiting above the plane of the accretion disk, ionizing the hydrogen and causing them to give off broad and narrow emission lines, depending on how fast the clouds move. If the accretion disk has an associated magnetic field, charged particles such as electrons will be accelerated and produce radio waves of synchrotron radiation. The radio waves will be emitted in two beams or jets, 180 degrees apart. A torus of dust surrounding the accretion disk will normally shield it from direct view. The high temperature will heat the dust and cause it to give off infrared radiation.

In this unified model, the type of AGN we see depends on the angle of view, the strength of the magnetic field (which may be related to the rotation of the central supermassive black hole), and how much material is being cannibalized at any given time. For example, if the AGN is viewed directly down one jet, a blazar would be seen, whereas if it were viewed at a slight angle, the object would appear as a QSO, and at a greater angle, it would be labeled a Seyfert. As the black hole cleans its plate (swallows most of the material orbiting close by), its source of energy dries up, and the AGN fades. Future feasts (perhaps caused by a small galaxy being eaten by its larger companion) can cause temporary flare-ups, but increasingly as time marches on, the black holes are put on meager diets. This is the current status of the SMBH at the center of the Milky Way. This suggests that our home galaxy was once a Seyfert galaxy or QSO, and observers on some distant planet billions of light years away may, at this very moment, be looking in our direction and seeing the Milky Way as it once was—a wondrous and violent AGN.

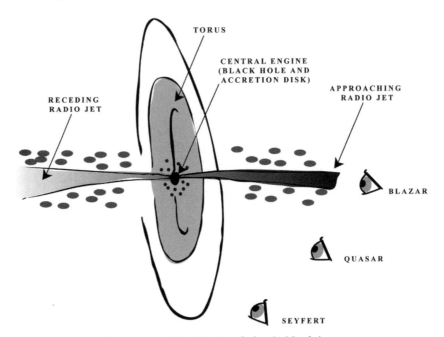

Figure 4.3 The Unified Model of Active Galactic Nuclei

DARK MATTER

Black holes are not the only case in cosmology where existence can only be verified through gravitational influence. In 1933, eccentric and brilliant Swiss astronomer Fritz Zwicky found that the mass of the Coma cluster, estimated from the luminosity of the galaxies it contained, was significantly less than the amount of mass needed to generate enough gravity to hold the cluster together. He posited the existence of some "dunkle materie," or *dark matter*. Three years later, Sinclair Smith found a similar result for the Virgo cluster, that the total mass of the cluster far exceeds the amount of mass visible in the individual galaxies. He suggested that the cluster might contain "internebular material ... in the form of great clouds of low luminosity" (1936, 30). Both studies were largely ignored by the scientists of the time, perhaps due to their focus on single galaxy clusters.

Evidence for dark matter in individual galaxies began to slowly accumulate several years later through the study of the rotation of spiral galaxies. In 1914, Vesto Slipher found that M31 and the Sombrero galaxy had a distinctive tilt to their spectral lines caused by the rotation of the galaxies (seen nearly edge on). This is very different from the rotation

Adriaan van Maanen had claimed to see in face-on spirals. As an edge-on spiral rotates, some of its stars are approaching us at varying speeds, whereas some are receding at similar speeds. This makes half of each spectral line blueshifted and the other half redshifted, causing it to appear tilted. The degree of tilt of different sections of each spectra line tells astronomers how fast stars in different parts of the galaxy are moving. A graph which plots rotational velocity against the distance from the center of the galaxy is called the galaxy's *rotation curve.*

An edge-on spiral resembles an over-easy fried egg, with the nuclear bulge being the yolk and the disk represented by the flat "white" of the egg. The halo of globular clusters is insignificant in this model. Observing the galaxy, it is clear that the majority of the mass should be contained in the nuclear bulge, and as we move farther out in the "white," we expect to encounter much less mass. This is similar to the solar system, where the vast majority of the mass is contained in the center, in the sun, and the planets form a relatively flat plane containing far less mass. If this model of the mass distribution for the spiral galaxy is correct, then spirals should rotate like the solar system, with the inner stars orbiting faster (as Mercury orbits the sun with a higher velocity) and the outer stars orbiting at much slower speeds (similar to Neptune). A graph of this type of motion, seen in Figure 4.4, is called a *Keplerian rotation curve* (named for Kepler's laws of planetary motion).

Horace Babcock's 1939 study of the rotation of M31 demonstrated a serious problem with this assumption, as he found that the stars in the outer parts of the galaxy orbited at nearly the same speed as those closer to the nucleus. Jan Oort obtained similar results for the spiral galaxy NGC 3115 the following year, and suggested that the galaxy contained either a large number of dim dwarf stars or some other unseen material. Babcock and Oort are given joint credit for recognizing the dark matter problem in individual galaxies.

Rotation curves for other galaxies, using visible spectra as well as the 21-cm radio line, continued to show that spiral galaxies do not follow Keplerian motion, but instead have flat rotation curves. The fact that stars orbiting near the visible edges of the galaxies orbit at nearly the same velocities as stars much closer to the nuclear bulge suggested that either our understanding of gravity is seriously suspect, or there is far more to a galaxy than meets the eye. Vera Rubin and her colleagues found that our own Milky Way has a flat rotation curve, a pattern that she would find in hundreds of spirals over several decades. Despite the accumulation of observational data suggesting some dark material dwelled in galaxies and galaxy clusters, the idea was not given wide credence until two

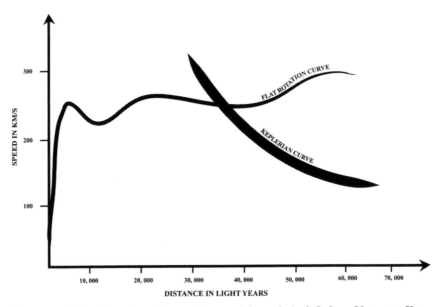

Figure 4.4 Flat Rotation Curve Observed for a Spiral Galaxy Versus a Keplerian Rotation Curve

theoreticians, Jeremiah Ostriker and James Peebles of Princeton University, found evidence in 1973 that spiral galaxies might become distorted or even fly apart unless they possess massive halos that extend beyond the visible galaxy.

The amount and distribution of dark matter within a galaxy can be found by modeling its observed rotation curve, and suggests that galaxies have roughly spherical halos of dark matter that extend beyond the visible edges of the galaxies and outweigh the visible stars, gas, and dust by several times. Determining the amount of dark matter in a galaxy cluster depends on accurately estimating the total mass of the cluster and subtracting off the mass contributed by stars and other visible constituencies. The first of three methods of cluster mass determination is that used by Zwicky, a study of the velocities of galaxies within a cluster. The second involves haloes of intracluster medium—hot gas discovered through its x-ray emissions. The hot gas would dissipate into space if it were not gravitationally bound to the cluster. The total mass of the cluster can therefore be estimated by calculating how much gravitational attraction would be needed to hold on to the observed gas, given its temperature.

The final method for determining the mass of a galaxy cluster involves another prediction of general relativity. In 1936 Einstein noted that if one star passed exactly in front of a more distant star, the image of the distant star would be gravitationally distorted or lensed into a ring of light. A less than perfect alignment would result in a more complex series of multiple images. Given the slight chance that two stars would align in this way, the prediction of *gravitational lensing* was largely ignored for several decades. That changed in 1979, when Dennis Walsh, R.F. Carswell, and R.J. Weymann discovered what seemed to be nearly identical twin quasars appearing very close to the image of a galaxy. Rather than claim that the two objects were instead a pair of unusually similar quasars, they suggested the radical solution that they were actually two images of a single quasar caused by gravitational lensing by the intervening galaxy. Since this original discovery, numerous examples of gravitational lensing have been found, including the so-called Einstein rings produced by the perfect alignment of galaxies and quasars.

Entire clusters of galaxies can also act as complex lenses, generating images of more distant objects as strange blue arcs of light. The amount of lensing can be used to construct a map of the mass distribution in the lensing object, whether it be an individual galaxy or a galaxy cluster, and has been a successful means of measuring dark matter distributions in clusters. The results of numerous studies are that galaxies only make up 10 percent of the total mass of a cluster, the hot intracluster gas contributes 10–25 percent, and the majority of the mass is found as dark matter. In August 2006, this method was used to find direct evidence of dark matter. The Bullet cluster is a small piece of a galaxy cluster which is ramming into another galaxy cluster. When galaxy clusters collide, the galaxies are too spread out to smack into each other, but the intracluster medium in the individual clusters collides, slows down, and heats up to millions of degrees, emitting x-rays. The Chandra x-ray satellite mapped the location of the hot gas in these merging clusters through its x-rays. Dark matter is rather aloof and doesn't play well with normal matter, so while the intracluster gas was busy colliding, the dark matter kept passing through, and separated out from the normal gas. The gravitational lensing tells us where all the mass is, the x-rays tell us where the normal matter, so by default the rest of the material is the dark matter.

LARGE-SCALE STRUCTURE

The existence of galaxy clusters led to the possibility of larger structures, now called superclusters. For example, the Local Group is a part of the Local Supercluster, a loose aggregation of galaxy clusters about 30 Mpc wide, centered on the Virgo cluster. Superclusters and larger structures are generally termed large-scale structure, and do not appear to represent features that are gravitationally bound to each other, in contrast to galaxy clusters. One possible exception is the infamous Great Attractor. In 1988, Sandra Faber, Alan Dressler, Donald Lynden-Bell, Roberto Terlevich, Roger Davies, Gary Wegner, and David Burstein unexpectedly found that our Local Supercluster is being pulled in a direction in space in which there did not appear to be anything large enough to so obviously disrupt the normal Hubble flow. The Seven Samurai, as the astronomers were later nicknamed, dubbed the unseen object the Great Attractor, and estimated that its mass must be that of a very large supercluster. The exact identity and properties of the supercluster (or superclusters) responsible for this motion are still uncertain, hampered by the fact that it appears to be located in the zone of avoidance and is therefore difficult to study.

The first hints of the mammoth scale of the largest structures came in 1981, when Robert Kirshner, Augustus Oemler, Paul Schechter, and Stephen Shectman discovered a huge "void" in the constellation Bootes. This zone of the sky, some 50 Mpc wide, was found to have many fewer galaxies than normal. It was expected that voids were not very common, and that matter was distributed rather smoothly, as predicted by the cosmological principle. The true extent of the voids and their intervening structures was found in 1985 with a *redshift survey* headed by Margaret Geller and John Huchra of the Harvard-Smithsonian Center for Astrophysics (CfA). In a redshift survey, the spectra for all galaxies in a small slice of the sky are electronically captured by a CCD. This slice would look similar to taking a piece of pizza and holding the pointed end up to your eye. The slice of the universe observed is very thin and stretches a larger distance side to side, causing the slice of space measured to be wedge-shaped. Additional wedges of the sky can be taken, either to the right or left or above and below in the sky, and by stacking slices a three-dimensional picture of the local universe emerges.

Their first slice measured galaxies out to about 100 Mpc, and the data was given to French graduate student Valérie de Lapparent to plot apparent location in the sky (right ascension or celestial longitude) versus redshift. To everyone's surprise, the galaxies were not evenly distributed, but were clumped together in a pattern similar to taking a slice through Swiss cheese or a sponge, with the galaxies being found on the surfaces of bubble-like structures some 30–60 Mpc wide. Several extended strings of galaxies, called walls, have been found in the CfA and other redshift surveys, and can stretch for several hundred Mpc. Note that they were plotting redshift, rather than distance, in this survey. From Hubble's law,

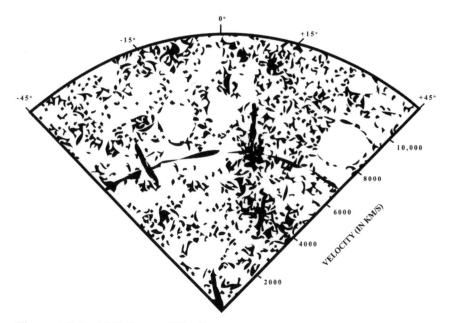

Figure 4.5 Redshift Survey "Slice"

velocity = distance $\times$ H, so knowing the redshift leads directly to the velocity, and therefore should give the distance. However, as we shall soon see, one must first separate out any peculiar velocities caused by the motion of galaxies in a cluster. The CfA plot (and most redshift surveys) fail to do this, resulting in some artificial features appearing in their maps. These structures are easily identified, as they look like fingers pointing directly at the origin of the map, as in Figure 4.5, and are called "Fingers of God."

This filamentary structure of matter in the universe seems to threaten the assumption of homogeneity presumed in the cosmological principle. However, it has been found that on larger scales, the universe does not become clumpier, but instead looks more uniform. This is similar to a beach. As seen from the ground, the dunes, rocks, and even individual sand grains give the beach a lumpy appearance. But as seen from an airplane, the beach looks smooth and featureless. As we have seen in this part of our journey through the cosmos, it may be homogeneous, but the universe is far from dull.

THE COSMIC DISTANCE LADDER

There is no single method which by itself allows astronomers to measure the distance from Earth to the farthest visible galaxies in a single step. Instead, there are a host of different methods, each of which is useful for a specific range of distances, sometimes collectively referred to as the *cosmic distance ladder*. By cross-checking different methods which work for overlapping ranges of distances, errors are minimized. Despite these good intentions, it should be noted that measuring the distances to the farthest objects is extremely difficult, and uncertainties (a polite way of saying errors) of 10–20 percent (or more) are sometimes encountered.

We begin literally in our own backyard, with the orbit of Earth. As described previously, the astronomical unit is a standard unit of measure in astronomy. The first (albeit imprecise) measurement of Earth's orbit was organized by Giovanni (Jean) Cassini in 1672. Mars appeared close to a fairly bright star, and simultaneous measurements were made from French Guyana and Paris. From the apparent difference in the positions measured (i.e. parallax), a distance of 86,000,000 miles was suggested. A more precise method was suggested by Edmund Halley, which utilized transits of Venus across the face of the sun, events which happen at a rate of about two per century. The transits of 1761, 1769, 1874, and 1882 were observed with these calculations in mind. Further refinements were possible in the 1960s, when radar waves were bounced off Mercury, Venus, and Mars. The speed of light multiplied by the time required for the round trip of the signals is twice the distance between Earth and each planet, and through geometry, the au can be derived

Once the au was known with good precision, the distances to nearby stars can be determined through trigonometric parallax, often abbreviated to parallax. This is the same effect that Aristotle could not see, and which he used as evidence against the motion of Earth. As we orbit the sun, we view stars from different perspectives. If observations are made of stars six months apart (on opposite sides of our orbit), as in Figure 4.6, nearer stars should be seen to shift their positions relative to farther (apparently fixed) stars. The only information needed is the size of Earth's orbit and the amount of the shift (an angle). In practice, however, the measurements are quite difficult. For even the nearest stars, the shift is a tiny fraction of a degree, measured in seconds of arc (where there are 3,600 sec of arc per degree). These small angles are generally smaller than the resolution of Earth-bound telescopes, as the atmosphere smears out the star images. It is therefore not surprising that successful parallax measurements were not made until the mid-1800s. Friedrich Georg von Struve published a parallax of 0.26 sec of arc for Vega (1837), Thomas Henderson found a parallax of Alpha Centauri

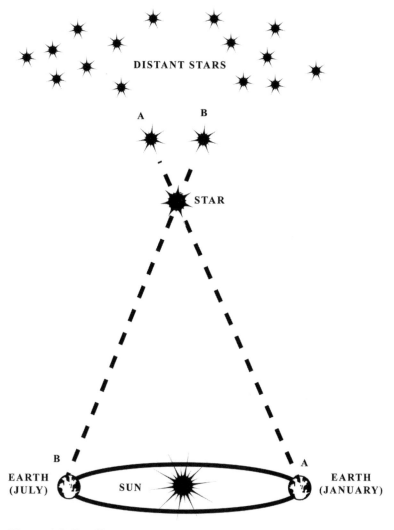

Figure 4.6 Parallax

of 1.26 sec of arc (1838), and Friedrich Wilhelm Bessel estimated a parallax of 0.31 sec of arc for 61 Cygni. All three values were too large, by factors ranging from 10 to100 percent (Webb 1999, 72). The parsec was defined as the distance of a star having a parallax of 1 sec of arc (being the contraction of "parallax-second"). In reality, there are no known stars that close to our own. The nearest star system, Alpha Centauri, has the largest possible parallax, and the modern value is only 0.74 sec of arc!

Given the difficulty in doing these measurements, it is not surprising that parallaxes for only seventeen stars were known by 1878 and a hundred by 1908. The seminal work on stellar properties, the *Yale Bright Star Catalog* (1952), listed fewer than 6000 parallaxes. In 1989, the European Space Agency launched the Hipparcos satellite (named after the ancient Greek astronomer), the first dedicated to the measurement of stellar parallaxes, positions, and motions. Following the completion of its data collection in 1993, an impressive catalog of 22,396 parallaxes (out to 90 pc) were measured to within 10 percent accuracy. Thousands of other parallaxes were measured with less certainty, and positional data on hundreds of thousands of stars were collected. Included in these stars were a number of Cepheid variables, which led to a further refinement of the period-luminosity relationship.

Stars move through space in three dimensions (their space velocity), but measurements are generally done in pieces. The radial velocity is calculated from the Doppler shift, while the motion across the plane of the sky is termed the tangential or transverse velocity. By combining the tangential and radial velocities through Pythagoras' theorem ($a^2 + b^2 = c^2$), the space velocity can be calculated. Although the various velocities are normally reported in km/sec, the apparent motion of a star across the sky is sometimes reported in seconds of arc per year, called the proper motion. The closer stars tend to have larger proper motions, just as a low flying plane seems to cover a large area of the sky faster than a higher flying plane. Proper motion was discovered by Edmund Halley, when he compared the current positions of the bright stars Sirius, Procyon, and Arcturus to those on ancient Greek star maps.

For nearby groups or clusters of stars, proper motion can be used as a method to determine distance. This moving cluster method is based on the principal that since the stars in the cluster are all moving like a flock of birds toward a common location, their proper motions will be in the same direction and if the arrows of their motion are extended, will all converge at some point in space. An example is five of the stars in the Big Dipper. Using trigonometry, the proper motion can be used to determine the distance to the cluster. Like parallax, this method is constrained to nearby groupings of stars. The most important "moving cluster" is the Hyades, a naked-eye group of stars that makes up most of the face of Taurus the Bull. At a distance of approximately 46 pc, Hipparcos was able to determine an accurate parallax distance for this cluster, which can be compared to that derived from the moving cluster method.

STANDARD CANDLES

At this point trigonometry and geometry have to be abandoned for less reliable distance measuring tools. At higher rungs of the cosmic distance ladder, the concept of *standard candles* becomes increasingly important. A standard candle is any astronomical object whose absolute magnitude can be estimated with some degree of confidence. The distance modulus formula is then used to calculate the distance. The most important example we've encountered so far is the period-luminosity relationship for Cepheid variables. The absolute magnitudes for standard candles should derive or calibrate their values from reliable methods (such as parallax), and wherever possible, a number of different standard candles should be used to estimate the distance to a given star cluster or galaxy in order to reduce error.

Besides problems with calibrating the absolute magnitudes of standard candles, extinction—the effect of intervening dust in making distant objects appear dimmer than they really are—adds to the error. Although astronomers have conscientiously tried to take extinction into account since the 1920s, there still remains some uncertainty. For very distant samples of objects, a third source of error creeps in. In 1920, Karl Gunnar Malmquist warned that a bias toward brighter members of any population of distant objects can lead to significant error. The *Malmquist bias* can be understood as follows. Consider a series of light bulbs of different wattages (10, 25, 60, and 75). The average brightness of the sample is 42.5W. If you observed the same sample from some distance away, you would not be able to see the 10W bulb because it is too dim. Therefore, based on the three bulbs you can see you would estimate the average wattage of the sample to be 53W. The farther and farther you are from a population of light sources, be they light bulbs, stars, or galaxies, the more dim members of that population you will overlook, and the brighter you will believe the average to be. This has significant implications for standard candles, where we need to know the absolute magnitude of a population of objects with certainty. The net result of all these issues is that with each higher rung of the cosmic distance ladder we (unwillingly) introduce greater error and uncertainty.

Let us return to the Hyades, whose stars' distances are known reliably through parallax and the moving cluster method. We can plot all 200+ stars on an HR diagram if we know their absolute magnitude and spectral class. Since we know the distance, we can turn the distance modulus formula around and calculate the absolute magnitude. The spectral class of the star is determined from the strength of various spectral lines. Recall that the width of the spectral lines places the star in its proper luminosity class (such as main sequence or giant). Therefore once we plot the positions of the Hyades' stars, we can calibrate the absolute magnitude for stars of a given spectral class and luminosity class. For example, the absolute magnitude of a main sequence F0 star is $+2.6$. while the absolute magnitude of a supergiant F0 star is -4.5. Of course, we have assumed that all supergiant F0 stars are exactly alike, which is not the case. But they are similar enough to use in this way. We can also use any other stars whose parallaxes are known to augment our calibrated HR diagram. The resulting diagram is then used to find the absolute magnitudes of other stars. If we know the star's spectral class and luminosity class from its spectrum, we can use the HR diagram to estimate its absolute magnitude. This is known as *spectroscopic parallax.*

Of course, a method is only as good as our calibration of the HR diagram. The concept of stellar populations becomes important here. Recall that stars of lower metallicity (population II stars) are dimmer than similar stars of population I. This was the origin of the subdwarfs, for example, as the population I equivalent of the main sequence (dwarfs). The problem is that of the stars close enough to have their parallaxes measured, only about 0.15 percent are population II stars. Therefore one of the purposes of Hipparcos was to measure enough subdwarfs to have a proper calibration for objects made of population II stars, such as globular clusters.

Of all standard candles, the most trusted are Cepheid variables. They are fairly bright and can be easily seen in galaxies of small and moderate distances. Since an individual galaxy can have many visible Cepheids, the average of many distance measurements can be taken and checked for consistency. Their distinctive variability makes them easy to pick out of a series of photographs and identified as Cepheids, rather than other kinds of variables. They remain in this variable stage for a long time (relative to the life of the average astronomer), so as technology improves, well-known Cepheids can be remeasured. Also comforting is the fact that the theoretical reasons for their variability are understood and can be modeled via computer. Therefore, once the period-luminosity relationship is calibrated (preferably using parallax), Cepheids are utilized with a high degree of confidence.

Cepheids, however, are not perfect tools. Since they are population I (younger) stars, they tend to be found in association with gas and dust, which leads to problems with extinction making them appear dimmer than they truly are. The fine-tuning effects of chemical composition (metallicity) on the period-luminosity relationship have not been resolved. As more distant galaxies are observed, the stars appear closer together, and it becomes more difficult to pick out the Cepheids. The current limit of useful observations of Cepheids is about 30 Mpc (30 million pc).

RR Lyrae stars are cousins to the Cepheids, and are also used as standard candles. Their absolute magnitude appears to be fairly constant, between 0.5 and 1.0, and the limited metallicity effects are believed to be fairly well understood. Because they are intrinsically much dimmer than Cepheids, they can only be used for nearby galaxies, within Local Group. Since RR Lyrae stars are population II (older) objects and Cepheids are population I objects, unless we take care to use both kinds of objects in the same galaxies and cross-calibrate, we run the risk of having two parallel distance ladders rather than one reinforced ladder.

Recall that both RR Lyrae and Cepheid variables occupy the instability strip on the HR diagram and represent stages in the old age and death of a star. Other stages in the late evolution of a star can be used as distance measuring tools. For example, the absolute magnitude of the tip of the red giant branch (TRGB) on the HR diagram can be used. If one plots an HR diagram for a globular cluster, the TRGB will show up as a sharp cusp, representing stars which are beginning to use helium as fuel. This is the brightest stage in the red giant evolution of a star. The absolute magnitude of the TRBG appears to be fairly consistent, with small (if any) metallicity effects, and can therefore be used as a standard candle.

When a low mass star completes its helium burning stage, it swells up to the asymptotic giant branch and eventually puffs off shells of material to form a planetary nebula. It has been found that although not all planetary nebulae have the same absolute magnitude, there is a well-defined distribution for the number of planetary nebulae at each absolute magnitude within a given galaxy. Since this planetary nebula luminosity function has a specific shape, it can be used to estimate the distance to a galaxy assuming enough planetary nebulae can be seen. Since they are found in both spiral and elliptical galaxies (unlike Cepheids), planetary nebulae can tie together distance-measuring techniques that are designed for a specific type of galaxy.

We have seen that the maximum brightness of novae has been used as a distance indicator since in the early twentieth century. However it was found that there was some dispersion in this value. More consistent indicators seemed to be the magnitude fifteen days after maximum and the rate of decline in brightness right after maximum. Because of the amount of data required to determine these values, novae have fallen out of favor as a standard candle. Fortunately when Knut Lundmark distinguished between novae and supernovae in the 1920s, a new standard candle was born.

As described previously, Type Ia supernovae (SNIa) are exploding white dwarfs, while Type II supernovae are exploding red supergiants. SNIa have become a well-trusted standard candle for several reasons. Because of their high luminosity (approximately −19 maximum absolute magnitude), they are easily visible over large distances (around 400 Mpc). This peak brightness seems fairly consistent from supernova to supernova. What dispersion there is can be adjusted for by looking at the shape of the light curve. Finally, the mechanism of the explosion is fairly well understood. However, searching for a SNIa is like looking for a needle in a haystack, and one never knows when or where one of these relatively rare events will occur. Another lingering problem, which will become of vital importance much later in our journey, is whether or not supernovae in the earlier ages of the universe are the same as those seen now. Type II supernovae are fainter and vary in their maximum brightness. While it is possible to estimate their distances through detailed calculations, they are not considered one of the favorite standard candles.

For galaxies too distant for individual stars to be seen, no matter how bright, the properties of the galaxy as a whole can be used to estimate distance. For example, in 1977 Brent Tully and Richard Fisher developed a relationship between the rotation of a spiral galaxy and its overall luminosity. Recall that unlike van Maanen's supposed observations, spiral galaxies are known to rotate, through their Doppler shifts. HI regions in the spiral arms of the galaxy orbit the center of the galaxies, as do the stars. Recall that HI regions emit distinctive 21-cm radio waves. However, as the galaxy rotates, some of the HI regions will be moving toward Earth (and have a slight blueshift relative to the center of the galaxy) while HI regions on the other side of the galaxy will be moving away from Earth (and have a similar slight redshift). The spread in wavelengths smears out the 21-cm line, and the amount of smearing (or width of the line) depends on the mass of the galaxy—the greater the mass, the greater the gravitational field, and the faster the rotation. Assuming that the more massive the galaxy, the more stars it has (and hence the more luminous it is), the width of the 21-cm line therefore allows one to estimate the overall absolute magnitude of the galaxy. It was found that because of the effect of dust in the spiral arms of the galaxies, it was more reliable to use the luminosity in IR rather than visible light, which is called the IRTF or Infrared Tully–Fisher relationship.

What about elliptical galaxies? Their stars do not neatly orbit the center of the galaxy, but instead have a more random motion. However, the average speed of the stars is also related to the total mass of the galaxy. In 1976 Sandra Faber (of the Seven Samurai) and Robert Jackson suggested that the dispersion in velocity—the spread in velocities from average—for stars near center of an elliptical galaxy can be used to estimate the mass and therefore the luminosity. It was unfortunately found that this Faber–Jackson relationship has a larger inherent uncertainty than the Tully–Fisher relationship. Marc Davis and Stanislav Djorgovski subsequently proposed that the velocity dispersion was better correlated to the diameter of a central region of the galaxy within which the apparent brightness was 20.75 magnitudes per square arc second. This fundamental plane relationship yields an uncertainty of 10–20 percent in distance for galaxies within a single galaxy cluster.

For galaxies so distant that not even these increasingly complex methods are effective, only one method remains. Recall that Hubble found that velocity = distance × H, where H is the Hubble constant. If we know H, then by measuring the velocity of a galaxy through its redshift, it seems simple enough to calculate its distance. That is, of course, assuming that the value of H is known with sufficient precision. That daunting task will be tackled after one intermediate topic—the question of the age of the universe.

THE AGE OF THE UNIVERSE

Although the age of the universe appears to be one of its most basic properties, arriving at a consistent and verifiable number has proven to be a challenge. There are two basic routes used to attack this problem—directly measuring the age of the universe (how long it has been expanding), or estimating the ages of the oldest objects visible in the universe and using this as a lower limit on the age. The former will be discussed in the next section, while the latter will be tackled here. The three types of objects whose ages have been used to set limits on the overall age of the universe are radioactive isotopes, white dwarfs, and globular clusters.

In 1929 Lord Rutherford introduced the use of radioactive atomic nuclei as a method to estimate the ages of rocks. When applied to the oldest earth rocks, moon rocks, and meteorites, this technique has been used to derive the currently accepted age of Earth and the solar system of 4.6 billion years. In the 1950s, seminal research by Margaret Burbidge, Geoffrey Burbidge, William Fowler, and Fred Hoyle (B^2FH) studied the production of elements in the universe, and opened the door for extending *radiometric dating* to the universe's oldest stars. The basic concept is as follows. An unstable atomic nucleus decays into another, lighter atomic nucleus at a specific rate. The original nucleus is called the parent, and the resulting nucleus, the daughter. The time required for half of a sample of radioactive nuclei to decay is called the *half-life*, and it is a measurable and constant property of that species of nucleus (or isotope). By comparing the relative amounts of parent and daughter in a sample, the age of the sample can be determined. For example, A decays into B with a half-life of 10 million years. If we find a rock with 50 percent A and 50 percent B we know that the rock is one half-life, or 10 million years, old. If we find a rock that has 25 percent A and 75 percent B, we know the rock is 2 half-lives, or 20 million years, old. The assumptions used here are that there was originally no B in the rock, and that B is stable.

Recall that the first generation of stars (population III) contained essentially only hydrogen and helium, and that the first atoms of heavier elements, including radioactive ones, were created in the old ages and deaths of these massive, short-lived stars. Therefore, by measuring the amounts of certain radioactive isotopes remaining in very old (but not first generation) stars, the ages of these stars can be estimated. One would then add to this age the lifespan of the first generation of stars (assumed to be of the order of a few million years) plus the time it took for that first generation of stars to form (a few hundred million years) to derive the age of the universe. This technique is sometimes termed *nuclear cosmochronometry*.

The two most useful isotopes are thorium-232 (which decays into lead-208 with a half-life of 14.1 billion years) and uranium-238 (which decays into lead-206 with a half-life of 4.5 billion years). Although the procedure seems straightforward, it presents serious difficulties to the astronomer. First of all, the spectral lines of thorium and uranium are weak, so getting an accurate measure of the relative abundance of these elements in a star is problematic. Once the current abundances are measured, one must know how much of these elements was originally present in the star, which means one must understand how they were produced in the first place. Despite these problems, there have been a few successes. Several very metal-poor stars have had their ages calculated as between 12 and 14 billion years.

White dwarfs, the corpses of low mass stars, can be used to estimate the age of the universe through their slow march toward oblivion. After a star reaches the white dwarf stage, it is no longer generating energy, so it slowly cools and dims at what is believed to be a predictable rate. By measuring the surface temperature of a white dwarf (through its spectrum), the time since its formation can be calculated, with the dimmest and coolest objects being the oldest. As with nuclear cosmochronometry, there are complications with this method. The rate of cooling depends on the composition of the white dwarf, particularly the relative amounts of carbon and oxygen. It is also important to note that the white dwarfs currently studied are in the disk of the Milky Way, which represents a younger sample of objects than the globular clusters. It is therefore expected that white dwarf ages will be significantly lower than those of the metal-poor stars used in nuclear cosmochronology. For example, a number of disk white dwarfs have been found with estimated ages of around 9 billion years.

By far the most widely trusted method for estimating the age of the universe using astronomical bodies involves globular clusters. Considered the oldest visible objects in our galaxy, they are relatively plentiful (over 150 known). The key is to understand stellar evolution, particularly how the HR diagram of a star cluster changes over time as its component stars evolve (with the more massive stars evolving faster than the less massive stars). The basic underlying assumptions are that the stars within the cluster were born at approximately the same time and have essentially the same chemical composition.

Let us examine the stars of a globular cluster soon after their birth. All of the stars will lie along the main sequence, with those of higher mass lying higher up (at hotter spectral classes and greater luminosities) along this narrow strip, and stars of lower mass occupying lower positions (cooler spectral classes and lower luminosities) on that same strip. Two billion years later, if we return to visit the same cluster, stars heavier than twice the mass of the sun (two solar masses) will have exhausted the hydrogen in their cores and left the main sequence for the red giant branch (or even death, in the case of the heaviest stars). Therefore the main sequence will no longer have any really hot and bright stars. The top of this clipped main sequence, seen in Figure 4.7 as the point where stars are just moving toward the red giant branch, is called the *main sequence turnoff point*, and is occupied by stars of exactly two solar masses. A visit 8 billion years later will show that the main sequence has retreated even more, and only stars with the same mass of the sun or less remain. The rest have become red giants or died. The main sequence turnoff point is occupied by these one solar mass stars.

The method of dating globular clusters is then as follows. An HR diagram of the stars in the cluster is plotted, and based on detailed models of stellar evolution, the position of the main sequence turnoff

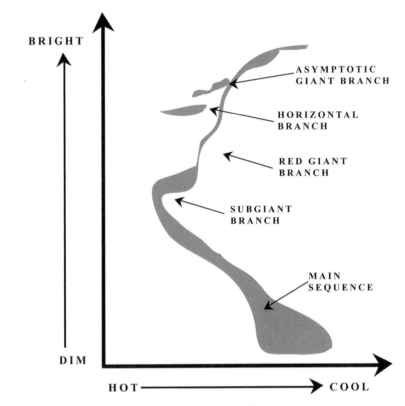

Figure 4.7 HR Diagram of a Globular Cluster

point is used to estimate the age. Implicit in this is the fact that the age determination is only as good as the understanding of stellar evolution. As computer models of stellar evolution have improved, the predicted ages for globular clusters have significantly decreased, from 16 billion years or more in the 1980s to less than 13 billion years in 2003. This decrease in estimates for the age of globular clusters occurred at a time of serious debate over the age of the universe, as will be described in the next section.

MEASURING THE HUBBLE CONSTANT

Since H, the Hubble constant, is the slope of the graph which plots the velocity and distance of galaxies, both those properties must be determined with some precision in order to have a well-defined and accurate value for H. The difficulties in determining distance have already been recounted. Although one might expect velocities to be more straightforward to measure, since they derive from the Doppler shift of a galaxy's spectrum, this is not the entire story. The Doppler shift is a measure of the galaxy's overall motion, toward or away from us. If a galaxy is isolated, with no nearby galaxies exerting a significant pull on it, then it is safe to assume that the entire motion of this field galaxy is due to the expansion of the universe, or Hubble flow, carrying the galaxy along. However, many galaxies occur in groups or clusters, where the individual members exert a significant gravitational pull on each other, and clusters exert a gravitational pull on other relatively close clusters. This portion of the motion, termed *peculiar motion or non-Hubble flow*, is responsible for creating the "Finger of God" effect seen in redshift surveys, and must be subtracted from the overall motion in order to determine the Hubble flow. Doing this accurately is much harder than it seems. For example, the galaxies in the nearest large cluster, the Virgo Cluster, exhibit a complex motion relative to each other, plus the overall motion of the cluster is being significantly affected by other neighboring clusters.

To avoid this problem, astronomers might only use field galaxies to plot a Hubble's law graph, or use the motion of a galaxy positioned at the center of gravity of a cluster (if one can really determine the center of gravity). The usual procedure is to use galaxies so far away that their peculiar motion is negligible compared with the expansion of the universe. But then we run into the problem that it is increasingly difficult to measure distances accurately the farther out we look. Another issue to keep in mind is that the value of H changes over time as the universe evolves. Because of this, the shape of the graph of Hubble's law is expected to deviate from a straight line when distant galaxies are plotted. The exact change predicted depends on the cosmological model used (for example, open, closed, or flat FRW models). Conversely, if a precise graph can be derived, the deviations from a straight line can be used to determine the geometry and evolution of the universe. We will revisit this important topic. Measurements of H using what might be considered our cosmic neighborhood determine the current value of H, more specifically called H_0 by cosmologists.

Since H_0 measures the rate of expansion of the universe, we can use it to derive the time since the universe began expanding. Roughly speaking, the age of the universe is the inverse of H_0 times a trillion $(1/H_0 \times 10^{12})$. This is a rough estimate, since H changes with time, and is actually most accurate for an open universe with no cosmological constant. If one assumes a flat universe with no cosmological constant, the age is two-thirds of this value $(2/3H_0 \times 10^{12})$. If one includes a cosmological constant, the universe can be much older than the value predicted from $1/H_0$. In 1936, Hubble himself was advocating a value of 530 km/sec/Mpc for the constant, which translates to a maximum age of the universe of 1.9 billion years. This is clearly younger than the age of Earth (as derived from rocks). As might be expected, this was deeply troubling to scientists, and was one of the motivations for Lemaître including a cosmological constant in his model of the universe. One could adjust the value of the cosmological constant such that there was good agreement between the age of the universe predicted by H_0 and the age of Earth.

Models with a cosmological constant fell out of favor, but a major breakthrough occurred in the 1950s, when Walter Baade discovered that there were two different types of Cepheids with different period-luminosity relationships. Distances to galaxies were recalculated and found to be greater, leading to a lower value of H_0. In 1956, Milton Humason, Nicholas Mayall, and Allan Sandage used over 800 galaxies to calculate H_0, which they found to be 180 km/sec/Mpc. This translates to a maximum age of the universe of 5.6 billion years, older than the age of the earth. Allan Sandage of Mount Wilson Observatory has devoted much of his career to refining measurements of H_0, and in 1958 published a further downward estimate of 75 km/sec/Mpc. He based this on a correction of several errors that had crept into previous estimates, including the misidentification of HII regions as bright stars, which had been erroneously used as standard candles by Hubble and others.

During the next four decades, there developed two "camps" in cosmology, one led by Allan Sandage and Gustav Tammann of the University of Basel, advocating for a lower value of H_0 (around 50 km/sec/Mpc), and another spearheaded by Gerard de Vaucouleurs (University of Texas) which argued for higher values around 100 km/sec/Mpc. The 100 percent difference in their numbers was based on what methods for measuring distances each group favored. Many observational astronomers leaned toward the higher values, while many theoreticians supported the lower ones, mainly due to the ages of the universe each predicted. For an open universe (with no cosmological constant), $H_0 = 50$ predicts the universe is 20 billion years old, while $H_0 = 100$ leads to an age of 10 billion years. As discussed earlier, the predicted ages for globular clusters stood at 16 billion or more years for several decades, in open contradiction to the $H_0 = 100$ value. The contradiction became significantly more serious in the 1980s with the development of the inflationary model of the early universe, which predicted that the universe was flat. This meant that the ages predicted by both $H_0 = 50$ and $H_0 = 100$ (13.3 billion and 6.7 billion, respectively) were too young to house the globular clusters.

Clearly a definitive study of H_0 was required. In 1986, a team of astronomers led by Marc Aaronson was awarded the Key Project on the Extragalactic Distance Scale on the then-unlaunched Hubble Space Telescope. The goal of the project was to observe Cepheids in targeted spiral galaxies (including those in the Virgo Cluster), and to use these measurements to estimate the distances to farther objects such as the Coma cluster in order to determine H_0 to within 10 percent uncertainty. Following the tragic death of Aaronson in a freak accident at Kitt Peak Observatory in 1987, Wendy Freedman took the reins of the project. In the wake of the Challenger shuttle tragedy, the launch of the HST was delayed from 1986 to 1990. Due to the optical problems of the telescope, major data collection was delayed until after special optics were installed on the cameras in December 1993.

In 1994, a paper by Freedman and her colleagues describing their initial results set the astronomical world spinning. Based on the measurements of Cepheids in M100, a galaxy in the heart of the Virgo Cluster, Freedman's team derived a value of 80 ± 17 km/sec/Mpc for H_0, and a value of 77 ± 16 km/sec/Mpc using the Coma Cluster. Understanding that these high values for H_0 were not consistent with the accepted ages of globular clusters in a flat (inflationary) universe, they openly predicted that "this 'age conflict' suggests that either the standard cosmological model needs to be revised, or present theories (or observations) bearing on stellar and galactic evolution may need to be reexamined" (Freedman, et al., 1994, 761).

The popular press was quick to smell the virtual blood. The March 1995 cover of *Discover* magazine was blazoned with "Crisis in the Cosmos," with the crowing pronouncement that "these days cosmology seems to be collapsing in on itself" (Flamsteed 1995, 66). The story quoted astronomer David Weinberg as saying "It would be premature to panic . . . but if these results are confirmed, we theorists will be in trouble" (Flamsteed 1995, 68). "Unraveling Universe" was the cover story of the March 6, 1995 issue of *Time*, where the "young whippersnapper" Wendy Freedman was set in dramatic opposition to "grumpy colleague" Allan Sandage. The article sensationalized the scientific debate, writing that

Nobody can say what the turmoil means—whether the intellectual edifice of modern cosmology is tottering on the edge of collapse or merely feeling growing pains as it works out a few kinks. "If you ask me," says astrophysicist Michael Turner . . . "either we're close to a breakthrough, or we're at our wits' end." (Lemonick and Nash 1995, 77)

As it turned out, the former proved to be true. Within several years, evidence for a cosmic repulsion akin to Einstein's cosmological constant began accumulating, which meant that the universe could be much older than the value predicted by H_0. At the same time, the ages of globular clusters were also revised, downward to about 12.6 billion years, which was within the timeframe predicted by an updated cosmological model including the cosmological constant. By the time Freedman's team published their final value of H_0 in 2001, 72 ± 8 km/sec/Mpc, the crisis had been resolved. Viewed through the detached lens of time, historians of science will dub the entire "cosmic age crisis" a textbook illustration of the self-correcting mechanism of science. It is also a reminder that scientists are human beings, and as such, can sometimes be extremely stubborn when faced with the demise of models, observations, or calculations they hold dear.

5

THE EVOLUTION OF
THE UNIVERSE

THE STANDARD MODEL OF PARTICLE PHYSICS

The Fundamental Forces and Particles

It may seem strange that a book whose subject is the largest thing that exists—namely the universe—should devote time to the study of the tiniest forms of matter. But consider the fact that the universe has not always been the unfathomable immensity it is today. The foundation of the big bang model is the concept that the universe emerged from a hot, dense state of being, in which the universe had more in common with high-energy particles than galaxies. Particle physics, the scientific study of the fundamental building blocks of nature, derives from quantum mechanics. Particle physicist Sidney Coleman once quipped that "if thousands of philosophers spent thousands of years searching for the strangest possible thing, they would never find anything as weird as quantum mechanics" (Randall 2005, 117). So far we have encountered several aspects of quantum mechanics—the wave–particle duality, Pauli exclusion principle and degeneracy pressure, and blackbody radiation. Another outgrowth is quantum field theory.

A *field* is some physical quantity which can be defined over a region of space-time. The best-known examples of fields are electric fields and magnetic fields (spoken of under the generic title of electromagnetic fields), and the gravitational field. Fields are invisible, but they affect matter and therefore can be measured and studied. For example, a charged particle such as an electron will move in a particular way in a magnetic field. The interactions of fields (with each other and with matter) are governed by field equations, such as the Einstein field equations of general relativity. In quantum field theory the fluctuations of the field are manifested as particles. For example, fluctuations in electric and magnetic fields produce photons (particles of light).

One of the strangest predictions of quantum mechanics involves the *Heisenberg uncertainty principle*. In classical physics it is possible to measure both the position of a car and its momentum at any instant as accurately as one's tools will allow. But consider an electron instead of a car, and the rules of quantum mechanics now apply. The very act of measuring the position of an electron (for example, by shining a light on it) changes its motion, preventing us from knowing both its position and momentum at the same time to absolute certainty. Another pair of variables so connected is energy and time. This means that it is possible for the universe to violate conservation of energy for a brief period of time, creating energy out of what most people consider the ultimate nothingness—the vacuum. This borrowed energy is used to create a particle-antiparticle virtual pair, called virtual because it only enjoys an ephemeral existence. In the blink of an eye, the pair annihilates, and in the process pays back the energy debt the universe accumulated from their creation. As strange as this process may seem, it is one of the most fundamental aspects of nature at the quantum level, and has measurable effects we can directly observe. For example, two parallel metal plates set in a vacuum will be attracted to each other via the Casimir Effect, not by the gravitational or electric forces, but because of the virtual pairs continually being created and annihilating between them. Virtual pairs also create a tiny shift in the energy levels of atoms, observable as the Lamb effect.

If quantum mechanics affects small particles of nature, it would be helpful to know what kinds of particles (and their associated antiparticles) exist. Molecules are made of atoms, and atoms are made of electrons, protons, and neutrons. Are these particles truly fundamental, or are they made of even smaller particles? Like a set of wooden Ukrainian nesting dolls, every time we peel off one layer of structure we find another, smaller level underneath. Have we found the most fundamental particles of nature? The search for this answer began in the 1950s and 1960s, when physicists collided atomic nuclei and other particles at higher and higher energies. They found that they could create an entire zoo of particles that resembled protons and neutrons but were much heavier. These *hadrons* (from the Greek for "thick") were suggested by physicist Murray Gell-Mann to be made of even smaller particles still, which he named *quarks*. This suggested that an entirely new language and classification system of particles was about to be born.

Particles in nature can be divided into two categories, *fermions* and *bosons*. Fermions obey the Pauli exclusion principle, while bosons, in the words of physicist Lisa Randall, "are like crocodiles—they prefer to pile on top of one another" (2005, 147). The particles that make up matter, such as electrons, protons, and neutrons, are fermions, while bosons, like the photon, are responsible for mediating the four fundamental forces of nature. These forces are gravity, electromagnetism, the *weak nuclear force*, and *strong nuclear force*. Hadrons are particles that interact with each other via the strong nuclear force (sometimes abbreviated as the strong force) because they are made of quarks. The strong force itself holds the quarks together to make the hadrons. Since this force is so strong that it overcomes the electric repulsion of quarks of the same charge, it earns its name. The strong force is 137 times stronger than the electromagnetic force, its nearest competitor in strength.

Hadrons can be divided into *baryons* (from the Greek for "heavy"), which include protons and neutrons, and mesons, which do not enter into our story. Since the vast majority of the mass of an atom is due to its neutrons and protons, what we consider "normal" matter is referred to as *baryonic matter*. Each baryon is composed of three quarks, held together by bosons called gluons. Therefore gluons are said to mediate the strong force. Quarks come in six varieties, called flavors, but these are not your normal chocolate, vanilla, and strawberry. Instead, particle physicists somewhat humorously dubbed them up, down, strange, charm, top, and bottom. Each quark has a fraction of the charge of the electron or proton, with the up, charm, and top quarks each having a charge of $+2/3$, while the others have a charge of $-1/3$. If you suddenly doubt your high school chemistry teacher, who taught you that particles could only have whole number charges, don't. Isolated quarks are not directly observable in nature. The trios of quarks in baryons have a total whole number charge. For example, a proton is a combination of two "up" quarks and one "down" quark (uud), which has a total charge of $2/3 + 2/3 - 1/3 = +1$. A neutron is one "up" and two "downs" (udd), with total charge $2/3 - 1/3 - 1/3 = 0$. The other flavors of quarks are believed to make up a miniscule amount of matter in the universe and will not bother us further.

Take a closer look at the proton for a moment. The two up quarks seem identical to us. But doesn't that violate the Pauli exclusion principle? There must be some property that makes one of the ups different from the other so they occupy a different quantum state in the proton. Particle physicists gave this property the name color, although it has nothing to do with how we normally think of the word. Of the three quarks in the proton, one is said to have "blue" color, another "red" and the third "green." The total proton is color neutral, as it has one quark of each color. Because of all the different possible color interactions, there are eight different types of gluons which mediate the strong force. The theory that describes how quarks interact to form hadrons is called *quantum chromodynamics*, or QCD. This theory has been very successful in explaining many aspects of the subatomic world, and in 2004 David J. Gross, H. David Politzer, and Frank Wilczek shared the Nobel Prize in Physics for their contributions to the development of QCD.

The weak nuclear force (or weak force for short) is about 10,000 times weaker than the strong force, and describes how quarks change flavor, as well as interactions involving a different class of fermions, called leptons (from the Greek for "light"). Just as there are six flavors of quarks, there are six different types of leptons, the most important being the electron and the electron neutrino (commonly called the neutrino, although in actuality there are three different types of neutrinos). Unlike the hadrons, the leptons are truly fundamental particles. To the best of our knowledge they cannot be subdivided into smaller pieces. Because its interactions involve both quarks and leptons, the weak force mediates the radioactive decay of atomic nuclei, and even the decay of the neutron itself. A free neutron (one not bound in a nucleus) falls apart, with a half-life of about 10 min, creating a proton, an electron, and the antiparticle of the electron neutrino (an antineutrino). At the quark level, one of the neutron's down quarks is converted into an up quark, with the release of an electron and the antineutrino. Since quarks have positive and negative charges, a weak force interaction can change charge (from positive to negative or negative to positive) or keep it the same. Because of this, there are three different bosons that mediate the weak force—W^+, W^-, and Z. This last particle has no charge and is similar to a photon, the particle that mediates the electromagnetic force, but unlike the photon, the Z has a nonzero mass. In 1984, Simon van der Meer and Carlo Rubbia shared the Nobel Prize in Physics for their experimental discovery of the W and Z particles.

Unifying the Forces

Four fundamental forces might seem a reasonable number, but physicists were intrigued by the possibility that this number could be an artifact of the state of the universe today. For example, before the mid-1800s, scientists believed that the electric and magnetic forces were separate entities. By 1868 James Clerk Maxwell had demonstrated that they are really two aspects of a more fundamental entity, the electromagnetic interaction. The quantum field theory description of the electromagnetic force is called *quantum electrodynamics*, or QED. Sin-Itiro Tomonaga, Julian Schwinger, and Richard P. Feynman shared the 1965 Nobel Prize in Physics for their work in developing QED. The similarities between the W and Z particles that mediate the weak force and the photon, which mediates the electromagnetic force, suggested that in high energy experiments (at a temperature of around 10^{15} K) these similarities would become even more overwhelming and the two forces would act as one—dubbed the *electroweak force*. Sheldon Glashow, Abdus Salam, and Steven Weinberg shared the 1979 Nobel Prize in Physics for developing the theoretical underpinnings of the electroweak theory. Experiments have verified the predictions of the electroweak theory (such as the properties of the W and Z particles), and the electroweak theory combined with quantum chromodynamics forms what is called the *Standard Model*. Although this model has been very successful in terms of experimental verification, there remain questions that suggest it is a work in progress. It relies on 19 arbitrary constants (such as the masses of the fundamental particles) whose values are not predicted by the theory itself, but rather the proper values are put in by hand. This arbitrariness of what should be fundamental constants of nature has led some to comment that the Standard Model may be successful, but it is certainly not elegant.

If the electromagnetic and weak forces unify at higher energies, could the same be possible with the strong force? It appears so. Although there is no one unique unification scheme, the class of solutions are called *grand unified theories* (or GUTs). Part of the problem is the unimaginable energy necessary to test the predictions of GUTs directly. In order to experimentally generate the necessary energy and temperature (around 10^{28} K), a particle accelerator ("atom smasher") over 20 pc long would be required (Guth 1997, 31). An advanced degree in particle physics isn't required to understand that this experiment will never be possible. How, then, can physicists test the various versions of GUTs and see which, if any, most properly describes the universe we live in? The key is the universe itself. According to the big bang model, the universe is only cold today (at an ambient temperature of around 3 K). When the universe was much younger, it was also much hotter, and the extreme energy necessary to unite the electroweak and strong forces was generated. It is no surprise that beginning in the late 1970s, particle physicists and cosmologists began joining forces to understand not only the tiniest particles of nature, but the infancy of the universe itself.

Although the final properties of GUTs are still being studied, there are three important predictions that could be experimentally verified. The first is the creation of particles called *magnetic monopoles*. These are particles that act like the north or south pole of a magnet, but isolated without the other pole. These particles have not yet been discovered. The second prediction is that the proton will very slowly decay, with a half-life much, much longer than the current age of the universe. Recall that the half-life is the time required for half of a sample to decay, so given a large enough sample, several protons should be seen decaying. The fact that protons have not yet been observed to decay means that their lifetime is very long indeed, and has actually been used to rule out some of the competing versions of GUTs.

The third prediction is the existence of particles called axions. They arise through a rather complex system of conservation laws that have been found to be slightly broken by the universe. Particle physicists have found that particles and antiparticles have slightly different properties (called the violation of charge invariance), that the laws of physics can differ when the spins of particles are reversed (violation of parity invariance) and that the laws of physics are even slightly different in some cases when reactions go forward versus backward (violation of time invariance). These violations are important, because they are believed to have allowed matter to dominate over antimatter in the very early evolution of the universe, a process dubbed *baryogenesis*. Axions are an outcome of some models of baryogenesis, and although they individually have small masses, they could exist in such large numbers that they could affect the evolution of the universe.

If three of the four fundamental forces can be unified in the high temperatures and energies of the early universe, why not all four? The problem is that the three forces combined in GUTs all play by the rules of quantum mechanics. The fourth fundamental force, gravity, is currently described by Einstein's general theory of relativity, and to say that quantum mechanics and general relativity don't play well together would be an understatement. Einstein failed in his attempt to unify the two forces best understood in his time, gravity and electromagnetism, for precisely this reason. A marriage between quantum mechanics and general relativity would result in a completely unified "theory of everything" (TOE), sometimes called *quantum gravity*. A number of different possible versions of quantum gravity are currently being explored by physicists, including *supergravity*, an outgrowth of the theory of *supersymmetry*. This is an attempt to achieve the ultimate unification by connecting the particles that mediate the forces (bosons) with those that make up matter (fermions). In supersymmetry (called SUSY for short), every boson is associated with a fermion, and vice versa. The still-unseen associated particles are called *supersymmetric partners*. For example, the electron, a fermion, would have a supersymmetric partner called the selectron, which would be a boson. Versions of SUSY that include gravity are called supergravity (or SUGRA), where gravity is thought to be mediated by a boson called the graviton. Its SUSY partner is an as yet unseen fermion called the gravitino. The temperature necessary to unify all four forces is phenomenal, estimated to be about 10^{32} K (a hundred thousand billion billion billion degrees).

Since the four forces of nature were unified at the high temperatures of the very early universe, they must have separated into the distinctive personalities we see as the universe cooled. The mechanism is termed *spontaneous symmetry breaking*. An everyday example of symmetry breaking can be found at a round dinner table. Typically the eight or so guests will sit down, look around, and be uncertain as to which bread plate or drinking glass is supposed to be theirs. There is a therefore perfect symmetry of the dishware on the table. However after an awkward moment or two, one bold person will reach out and grab a glass, or subtly drag one of the bread plates closer to his main plate, thus breaking the symmetry for the entire table. If the person chose the glass to his left, so must everyone else at the table. In quantum field theory, symmetry preserving states are unstable, as if the symmetry wanted to be broken.

The actual breaking of the symmetry is done by a theoretical particle called the *Higgs particle*, proposed by Peter Higgs in 1964 to explain why particles such as electrons have mass. It is an example of a scalar particle, which has no spin. Since particles and fields are interchangeable through the wave-particle duality, there also exists a Higgs field, which is a scalar field. This field is simpler than the electric or magnetic field, because only the value of the scalar field at a point, not its direction, is measured. For example, a weather map that plots temperatures represents a scalar field, whereas a weather map that plots wind speed and direction does not. In grand unified theories there are multiple Higgs fields, the actual number varying from model to model. The symmetry of the model is preserved when all Higgs fields are zero, but as soon as least one of them takes a value other than zero, the symmetry is broken. The electroweak theory uses a different Higgs field in its symmetry breaking. Although the Higgs particle (and its related field) has so far eluded detection, most physicists expect it to be only a matter of time (and higher energy experiments). Although the Standard Model does not require the existence of the Higgs particle from first principle, without it the model is only mathematically consistent if all the fundamental particles are massless and travel at the speed of light (like the photon). Since this is clearly not the case in our universe, the Higgs particle (or something very similar to it) is needed to make most of the fundamental particles massive rather than massless.

THE FIRST FIFTEEN MINUTES

If we follow the expansion of the universe backward in time, we seem to be driven to a singularity, where the entire universe is compacted into a mathematical point of infinite density and zero volume, and where the equations of general relativity break down. This seemingly ridiculous prediction is a symptom of our lack of a unified theory of quantum mechanics and general relativity. Once such a "theory of everything" is developed, it is expected that the singular birth of the universe will be avoided. This also means that we currently lack the information to trace the history of the universe all the way back to the absolute beginning. Just how far back can we reliably go? This is determined by the scale of nature at which quantum effects cannot be ignored in gravitational calculations, which in turn is determined by the values of three fundamental constants of nature—Newton's gravitational constant (which sets the relative strength of gravity), the speed of light (which sets the maximum speed limit), and the Planck constant (which sets the minimum uncertainty in measurements). Combining these three constants, it is found that the scale of quantum gravity is measured using Planck units—the Planck length (10^{-35} m), the Planck time (10^{-43} seconds), and the Planck mass (10^{-5} g). The Planck length can be thought of as the size of the quantum mechanical "texture" of space-time, the Planck time as the earliest time in the universe we can discuss without quantum gravity, and the Planck mass is the smallest possible mass of a black hole.

Our tour of the history of the universe begins at the Planck time, 10^{-43} sec after the event that spawned the big bang. Before this time, in the *Planck era*, all four fundamental forces were one unified superforce, whose details await a successful theory of quantum gravity. At the end of the Planck era, the universe underwent a spontaneous symmetry breaking, and the gravitational force became separate and distinct, destined to remain so for the rest of time. The next period is called the GUT era, which can be described by grand unified theories. The universe is a soup of quarks, leptons, and photons interacting in madly energetic reactions that we cannot replicate in our laboratories. It is in this era that baryogenesis takes place, and matter gets the upper hand over antimatter. Pioneering theoretical work on baryogenesis was done in 1967 by Russian physicist Andrei Sakharov. An advocate of human rights and nuclear weapon opponent, he received the Nobel Peace Prize in 1975. Ironically, he was one of the scientists who helped the Soviet government develop the hydrogen bomb. At 10^{-35} sec (10^{28} K) another symmetry breaking event takes place, and the strong force takes on a separate life of its own, leaving the electroweak force behind. Leptons do not "feel" this newly distinct strong force, so they and quarks can no longer interact as equal partners as they had before. Another important event is theorized to have occurred at this important transition, namely *inflation*, which will be a major theme in a later section.

During this Electroweak era, the temperature cools to 10^{15} K (a thousand trillion), and when this temperature is reached (10^{-11} sec), the electromagnetic and weak forces separate from each other, establishing the four fundamental forces we experience today. It is amazing to consider that no new forces have established themselves after the first ten trillionths of a second of the universe's history. The next epoch is called the Hadron era, in which quarks form hadrons and just as quickly the hadrons are broken back up into quarks. Only at the end of this era (10^{-5} sec), when the universe has cooled to a mere 10^{10} K (ten billion Kelvin), can the quarks remain confined into hadrons. This results in the birth of protons and neutrons.

With this phase completed, the stage is finally set for the formation of the first elements. During this Lepton era, as in the previous epochs, photons (radiation) and matter were in thermal equilibrium. This means that because of the ferocious high-energy interactions, the temperature of the radiation was the same as the temperature of the matter. At temperatures higher than about 10^{10} K, the slight difference in mass between the heavier neutrons and the lighter protons is ignored by weak force interactions, and protons and neutrons turn into each other in equal numbers. For example, a proton and an electron can combine to create a neutron and an electron neutrino, or vice versa, with equal probability. When the universe cools below this critical temperature, the neutrinos decouple, meaning they fall out of equilibrium with the other particles and become free. This background of relic neutrinos fills the universe. With the neutrinos essentially out of the game by a few seconds after the beginning of the universe, neutrons were doomed to decay, with their characteristic half-life of about 10 min. But fortunately long before all the neutrons in the universe could decay, another critical temperature, 10^9 K, was reached, at about 3 min, and the Lepton era was completed.

In order to form atomic nuclei more complex than simple hydrogen (which is, after all, merely a single proton), the first stage is to combine a proton and a neutron to form a deuteron, a nucleus of deuterium (heavy hydrogen). However, at temperatures above a billion degrees, deuterons are torn apart by collisions with high-energy photons as soon as they are formed. This deuterium bottleneck holds up the creation of atomic nuclei until about 3 min into the history of the universe. Once the bottleneck is breeched, neutrons and protons rapidly combine to form in succession deuterons, nuclei of helium-3 (one proton and two neutrons), helium-3 (two protons and one neutron), and finally helium-4 or alpha particles (two protons and two neutrons). This continues until all the free neutrons are used up, and once the neutrons are bound into nuclei, they do not decay but are stable. This *primordial or big bang nucleosynthesis* fixes the overall chemistry of the universe at roughly 75 percent hydrogen and 25 percent helium, with tiny amounts of deuterium and helium-3.

Recall that there are no stable nuclei with atomic mass of five, so the few nuclei that jump the gap end up producing a small amount of lithium-7. The second mass gap, at eight, is much harder to jump, and primordial nucleosynthesis ends about 15 min after it begins. The universe must wait for the creation of the first generation of stars for further modifications in its chemistry.

THE DARK AGE

After the end of this Nucleosynthesis era, the universe entered a long stretch of relative quiet. The temperature of the universe, as determined by the sea of photons it contained, cooled as the universe expanded further. It is at this period of time that the observed ratio of photons to matter particles, a whopping two billion to one, was fixed. Thus far, the evolution of the universe had been dominated by the effects of radiation, leading to the generic term of Radiation-dominated universe for this portion of the universe's life.

As the universe expanded, the energy density of photons eventually became less than the energy density of matter, and the universe became Matter-dominated. Shortly afterward, about 300,000 years after the first instant of time, the universe underwent another vital transition, that of *recombination*. A rather confusing choice of words, this refers to the fact that the temperature had finally cooled enough (to about 4,500 K) to allow electrons to be bound to the hydrogen and helium nuclei for the first time and the first stable atoms were formed. With the electrons safely bound into atoms, there was nothing to scatter the photons, and they flew through the universe unimpeded. Previously the universe had been shrouded in an opaque fog of scattered photons, but it now became transparent. These photons continue to flood the universe today as the cosmic microwave background. Over the intervening billions of years, their wavelength has been dramatically redshifted and their temperature has dropped to its current value of about 3 K.

While radiation evolved on its own after recombination, matter began the first steps toward accumulating into the structures we see today. In keeping with the cosmological principle, matter was more or less uniformly distributed throughout the universe, but not perfectly so. Regions of matter with slightly higher than average density tended to clump together under the influence of gravity, eventually leading to the first generation of stars. Before this occurs, the universe is plunged into a temporary but utter darkness, as the CMB is redshifted from visible to infrared wavelengths. This *dark age* lasted from a half million to about 200 million years after the beginning of the universe. When the first generation of stars, called population III, turned on, the universe was again lit.

Discerning the details of the birth of this first generation of stars has been a challenge for astrophysicists, because they were made from the raw materials produced by primordial nucleosynthesis—nearly pure hydrogen and helium with virtually no "metals." One of the problems is that in the formation of normal stars (those which contain metals), metals and dust grains cool the nebula out of which the star forms, allowing gravity to win out over the outward pressure of radiation and collapse the material into a protostar. Molecular hydrogen (H_2) is believed to have acted as the cooling (heat-dissipating) agent in population III star formation and is very inefficient.

Another important difference between population III and population I and II stars is that they tend to form with unusually large masses—100 times the mass of the sun or more. These supermassive stars consume their fuel at an astounding rate, shining with a luminosity over a million times that of the sun. Because they lack heavy elements such as carbon, they cannot use the carbon cycle to fuse hydrogen but are limited to the proton–proton cycle despite their high core temperature. They also lack the necessary materials to make carbon-based life-forms or even solid planets like the earth. They could have had gas giants like Jupiter, but these unfortunate planets would have been roasted by their peculiar stars' unusually high surface temperature—nearly 110,000 K.

After a brief lifespan of only a million years, these first stars would have consumed all their fuel and begun to die. But their deaths may possibly have differed greatly from those of future generations. Stars with a final mass less than 140 solar masses or greater than 260 solar masses would have died as black holes, trapping most of their chemically enriched mass in an unusable form. Population III stars with masses in the 140–260 solar mass window are theorized to die by a process called *pair-instability supernovae* (PISN). In these stars, when oxygen fusion begins the temperature is high enough for photons to create electron–positron pairs. A positron is the antiparticle to the electron, and has a positive charge, hence its name. This reduces the outward pressure which normally counteracts gravity, and the star suffers a precipitous implosion which "triggers the explosive burning of the remaining nuclear fuel in the star, leading to its complete and utter destruction" (Bromm 2003, 30). No corpse, either neutron star or black hole, is left behind. This means that all the chemically enriched material contained within the star is blown off into space, enriching the interstellar medium.

Once the interstellar medium reaches a critical metallicity, normal star formation begins, and stars of only a few solar masses (or lighter) are possible. The Wilkinson Microwave Anisotropy Probe (WMAP), a space telescope that makes very precise measurements in slight variations in the cosmic microwave background in different directions of the sky, found evidence of this first generation of stars in 2006. These hot, massive stars ionized the interstellar medium, ripping electrons from their host atoms. The photons of the cosmic microwave background interacted with these electrons and were polarized in a unique and observable way. This reionization of the ISM and resulting radiation pressure made it harder for gravity to collapse clumps to form structures. Star formation probably occurred at a slower rate, but more importantly this might have affected the formation of larger structures such as globular clusters and whole galaxies.

Although the only evidence for these first stars has so far been indirect, in their deaths they may have provided direct signs of their existence. It has been suggested that those population III stars that had masses either too low or too high to die as PISN would emit tremendous bursts of gamma-rays before forming black holes. The NASA Swift satellite is currently monitoring gamma-ray bursts, and it is hoped that many candidate signals from the distant past will be observed over the next few years.

THE FATE OF THE UNIVERSE

The ignition of the first population III stars heralded the beginning of what astrophysicists Fred Adams and Greg Laughlin have dubbed the Stelliferous ("Star-filled") era. It will end when the last stars burn out. The most frugal of all stars, low mass red dwarfs, have an estimated lifespan of at least 10^{13} years, some 1000 times the total lifespan of the sun. It is also estimated that when the universe is between 10^{12} to 10^{14} years old, star formation will stop, as all available hydrogen gas dense enough to collapse to form stars will have been used up. Therefore not much longer than 10^{14} years after the first stars were born, the last ones will die. The fate of galaxies near the end of the Stelliferous era is no less gloomy. Individual galaxies within clusters will merge, as our Milky Way may do with the Andromeda Galaxy in an estimated 6 billion years, creating one large mass of dying stars.

The death of the last red dwarfs brings us to the start of the Degenerate era, in which the universe is dominated by stellar corpses—white dwarfs, neutron stars, and black holes. The name comes from the fact that white dwarfs and neutron stars are held up from further collapse by the degenerate pressure created by the Pauli exclusion principle. The universe is locked in a dim twilight, occasionally punctuated by the collision of two brown dwarfs, igniting a temporary period of nuclear fusion, or the supernova explosion of coalescing white dwarfs. The Degenerate era comes to an end about 10^{39} years after the universe begins, the actual timing determined by the decay half-life of the proton. Once all the protons contained within white dwarfs and neutron stars have decayed (into positrons, neutrinos and other particles), the only remaining stellar corpses are black holes.

In the Black Hole era, the only physical process of importance is the slow evaporation of black holes predicted by Stephen Hawking. The amount of time it takes for a black hole to decay into elementary particles increases as the mass increases. The largest theorized black holes would contain the mass of an entire galaxy, and would take approximately 10^{100} years to self-destruct. Note that the number 10^{100} is called a googol, not to be confused with the search engine of a similar name. The demise of the black holes marks the end of the Black Hole era. Time marches onward, as the universe enters the Dark era, with space-time utterly empty with the exception of widely scattered photons, neutrinos, electrons and positrons. As it was before the first generation of stars, the universe is a dark and lonely place, certainly not a comforting thought.

When poet Robert Frost pondered whether the universe would succumb to fire or ice, even his creative mind could not have imagined a fate so bereft of hope. A more hopeful future appears in the classic short story, "The Last Question," where Isaac Asimov invokes a somewhat religious solution to this scientific inevitability.

INFLATIONARY COSMOLOGY

Problems with the Big Bang

Although the big bang model was a success in fitting with the cosmological observations of the 1960s, a small number of nagging problems subsequently began to emerge. Four of the most important are the flatness problem, the horizon problem, the density perturbation problem, and the magnetic monopole problem. The first to be discussed was the flatness problem, first posed in a lecture by Robert Dicke of Princeton in 1969. It asks why the current density of material in the universe is so close to the critical density needed for a flat Friedmann–Robertson–Walker model. In other words, why is Ω so close to one? Based on simple counts of the visible matter in the universe, one gets an ultimate lower limit of $\Omega \sim 0.015$, and taking into account the dark matter contained in galaxies and galaxy clusters, this limit climbs to 0.2–0.3. While these numbers may seem vastly different from 1, consider the following—why are they not 0.000001 or 50,000? In terms of order of magnitude, 0.015, and especially 0.2, are rather close to the special value of one.

The importance of this seeming coincidence becomes clearer when considering what Ω itself determines. If $\Omega > 1$, the universe is closed because it is "too dense" and is doomed to recollapse, assuming there is no cosmological constant. If $\Omega < 1$, there is insufficient density to close the universe, so it is open, and it will expand forever at an ever-slowing rate. The case of $\Omega = 1$, a flat universe, is a finely tuned value, and is often compared to balancing a pencil on its point. In the case of a balancing pencil, any slight deviation or fluctuation to one side or the other will send it toppling over. Likewise, any slight wiggle from the absolute value $\Omega = 1$ will mushroom over time, driving Ω farther away from one as the universe evolves. In order for Ω to be as close to 1 as we observe today, it would have originally had to have been precisely equal to 1 to within 1 part in 10^{49} or better. While it is certainly possible that the universe was created with exactly this special value, there is no reason in the classic big bang model why it should have chosen such an inherently unstable value, and some scientists regard the reason "because it just happened to be that way" to be artificial at best.

Another artificial-appearing observation of the universe involves the cosmological principle. On average, the universe should be homogeneous and isotopic, not only in terms of its matter, but also the cosmic microwave background. The key words are "on average." Measurements of the CMB, the energy fingerprint of the hot early universe, have consistently shown that it is extraordinarily smooth, to within a few parts in 100,000. It is possible that the initial state of the big bang was absolutely uniform in temperature to this amount (i.e., it was created in remarkable thermal equilibrium), but exceedingly unlikely. By comparison, consider the water coming out of the tap in your bathtub. The water has natural fluctuations in temperature, sometimes running hotter, sometimes cooler. Before stepping into the tub, the wise bather either stirs the water to help it equalize in temperature, or waits a few minutes for equilibrium to be reached naturally. Now imagine how difficult this would be if the bathtub were expanding! In order for the different parts of the universe to reach equilibrium, they would have needed to communicate with each other far faster than the speed of light.

This is also called the horizon problem, for the following reason. As you look up into the night sky, you could (with a telescope of the greatest theoretical power), see as far as 13 billion light years in each direction, if the universe is 13 billion years old. This is because you can only see light which has had sufficient time to reach your eye. A year from now, you could see 13 billion + 1 light years in each direction, and so on. This means that observable patches of the universe on opposite parts of your sky are separated by 26 billion light years, or it would take 26 billion years for a light signal to travel from one side of your "horizon" to the other. How, then, can the CMB on opposite sides of the sky be at the same temperature to within a few parts in 100,000? One might think it would be possible for information to have been freely exchanged when the universe was much smaller, but it was initially expanding so fast that it contains at least 10^{83} causally disconnected regions that have still not had sufficient time to communicate.

Despite the remarkable smoothness of the CMB, it is not completely smooth, otherwise we would not exist today. The minute fluctuations in temperature now observed are a reflection of the fact that, while homogeneous on the largest scales, matter clumps at smaller scales, in the form of galaxies and clusters of galaxies. There must have existed "seeds" in the early universe from which these structures grew—namely fluctuations or perturbations in the density of matter. The original big bang model is silent on the source of these all-important deviations from perfect homogeneity, sometimes called the density perturbation problem.

Theories that seek to describe the early eras of the universe, namely GUTs, are the source of the final problem. Recall that they predict the existence of massive particles named magnetic monopoles. The problem is the high number predicted to exist—a trillion times more (by mass) than all the observed matter in the universe! Increasing the density of the universe by such a mammoth amount would not only make it closed (dooming it to a fiery "big crunch"), but it would have slowed the expansion of the universe to the rate currently observed in about 30,000 years. Given that the age of the universe is now estimated to be over 13 billion years, there is clearly a serious problem between theory and observation. Not only is the predicted density of magnetic monopoles catastrophic, but not a single magnetic monopole has yet been observed. In the words of MIT physicist Alan Guth, "like the unicorn, the monopole has continued to fascinate the human mind despite the absence of confirmed observations. The monopole, however, has a much better chance of actually existing" (1997, 30). It was the search to find a solution to this *magnetic monopole problem* that led Guth to develop a model which was actually capable of solving all four problems.

Old Inflation

In 1981, Guth was researching the details of spontaneous symmetry breaking in the early universe, specifically at the end of the GUT era (10^{-35} seconds into the history of the universe) when the strong force separated from the electroweak force. Specifically, he investigated the possibility that when the transition temperature at which the symmetry breaking should occur was reached, the universe failed to immediately go through the transition. A similar process can occur when water is cooled below its freezing point yet remains liquid, as in the case of freezing rain. This is termed *supercooling*. To understand what happens if the universe is in a supercooled state, we need to revisit the concept of Higgs fields.

It is simplest to picture a "toy" model where there are two Higgs fields, which can be represented pictorially like a chessboard. The center of the chessboard represents both Higgs fields having a value of zero. A third dimension—height—can now be added to our picture, representing the energy density of space-time. The lowest possible value of this energy density (remaining flat on the chessboard) is called the true vacuum.

If this is called the "true" vacuum, it suggests that there must be a *false vacuum*. Picture a Mexican sombrero placed on our chessboard. It has a slight dimple or dent at its top. For a bug crawling around the top of the hat, the bottom of the dimple looks like the lowest possible point. Scientists say that it represents a local minimum in the height of the hat. However, a wider view shows that this is not the true lowest point in the hat—the round trough in the brim of the hat (where it touches the chessboard) has that honor. This is the global minimum of the hat. The dent in the top of the hat is the false minimum or false vacuum, whereas the trough in the brim is the true minimum or true vacuum. Note that the dent at the center is a state of symmetry, as all the Higgs fields are zero, whereas the brim of the hat is a state of broken symmetry, as the Higgs fields are not zero. The surface of the hat represents how the energy density depends on the values of the Higgs fields.

Now consider a small ball rolling on the surface of the hat. The act of rolling from the top to the bottom represents a phase transition, from a state of preserved symmetry to a state of broken symmetry. It is possible that the ball could get "stuck" in the false vacuum in the top of the hat and be unable to complete the transition to the broken symmetry state. But recall that nature "wants" to break the symmetry, so there is a price to be paid for being caught in the false vacuum. If the ball were a bubble of the early universe it would be in a *supercooled* state. While stuck in the false vacuum, the bubble would be under the influence of a repulsive force, and it would inflate exponentially, growing 10^{25} times bigger in the blink of an eye. Such an exponentially expanding universe is described by the de Sitter solution to the Einstein field equations. The universe inflates so much that the density of matter declines to essentially zero, making the matterless de Sitter model a good approximation. However, we obviously do not currently live in a de Sitter universe, so the bubble must somehow get out of the false vacuum state and reached the broken symmetry state of the true vacuum.

In Guth's original inflationary model, the bubbles cannot get out of the false vacuum "dimple" by rolling up and over the hump, because they don't have enough energy to do so. Instead, as shown in Figure 5.1, they "tunnel" out *through* the hump. This is clearly impossible by the classical laws of physics, but it is common enough in the strange world of quantum mechanics. *Quantum tunneling* is the basis of such technological advances as the tunnel diode in electronics and the scanning tunneling microscope. In this so-called old inflation, the universe tunneled out one bubble at a time, and somehow the bubbles collided and recombined on the "other side"—in the broken symmetry state. However, as Stephen

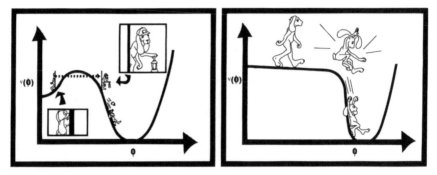

Figure 5.1 Old vs. New Inflation

Hawking and others later showed, the bubbles would never actually merge. Thus old inflation suffered from what is called the "graceful exit problem"—there is no realistic way to end the inflationary phase.

New Inflation

The new inflation model, independently proposed by Andreas Albrecht and Paul Steinhardt of the University of Pennsylvania, and Russian physicist Andrei Linde, avoided this problem by not having bubbles try and reconnect. Instead, the entire visible universe was spawned from a single bubble of false vacuum. The mechanism for leaving the false vacuum state was different as well. Instead of an "energy barrier" which had to be tunneled through, the shape of the "hat" included a long, flat region, giving this model its sometime name of the "slow rollover." Quantum fluctuations would start the ball rolling, in a literal sense, slowly driving the bubble off of the false vacuum plateau, inflating as it rolled. When the ball reaches the bottom of the hill (the brim of the sombrero), it would oscillate back and forth, reheating the supercooled universe to the level expected by the big bang model. Some of the energy would be converted into particles, repopulating the now-empty universe.

Inflation was developed in response to problems with the original big bang model. How successful was it in solving those problems? At the end of the inflationary period, the density of any particles created beforehand is diluted to essentially zero by the huge increase in the volume of the universe, thus solving the magnetic monopole problem. This also explains why there has as of yet been no experimental observation of magnetic monopoles—the nearest one could now be many parsecs away!

The incredible expansion of the universe also solves the smoothness problem in much the same way. Any inhomogeneities in the CMB which existed *prior to* inflation would be smoothed out, just as wrinkles in fabric (or skin) disappear when it is stretched. Another way of looking at the problem is that it is no longer assumed that the entire visible universe arose from a state of thermal equilibrium, only that what we see was born from a tiny patch of the initial big bang which, in itself, had achieved a constant temperature over its small volume. A solution to the flatness problem follows close behind. Any large-scale curvature to space-time which was created by the initial big bang would be "flattened out," as the surface of a sphere (like Earth) looks flat when it becomes large enough.

The density perturbation problem was the most complicated to solve, and it was finally realized that the Higgs field itself is not the field responsible for inflation. Instead, there must be some other weakly coupled field, dubbed the inflaton field, which drives inflation. The exponential expansion of the universe ends when the phase transition is made from false vacuum to true vacuum, or, in pictorial terms, the inflaton field "ball" reaches the steep part of the energy hill. Due to quantum fluctuations, these fluctuations would not be the same everywhere. These fluctuations would be the source of small fluctuations in the density of matter in the early universe, which would in turn act as the seeds for the formation of larger structures such as galaxies.

A common misconception about inflation is that it replaces the big bang. It is actually a modification of the classic big bang model, developed to answer several problems and nagging questions. Although the exponential expansion which gives the theory its name only lasted for an infinitesimally small fraction of a second, its effects are visible for the entire history of the universe. After inflation, the universe evolves as predicted by the big bang model. Cosmologist David Schramm called inflation "pretty," noting that "It really gave a boost to this interface of cosmology and particle physics, capturing some of the most exciting aspects of both" (Lightman and Brawer 1990, 445).

Eternal Inflation

Inflation is also not a monolithic theory, but a paradigm or class of theories that can derive from different sources and make measurably different predictions for the universe we inhabit. For example, the universe could have entered an inflationary state by supercooling from high temperatures (as recounted above), or, as in the research of Alexander Vilenkin of Tufts University, the result from tunneling from "nothing" directly into a false vacuum state. Here "nothing" literally refers to a state of no classical space or time. A third possibility, developed by

Andrei Linde, is called chaotic inflation. In this version of inflation, the inflaton field can find itself in a variety of different energy levels at different locations due to random fluctuations. In some places, it happens to be in a state of higher energy, which acts like a false vacuum state and that region of space-time would inflate. In chaotic inflation, different universes are born from this initial state at different times and individually evolve into their own separate, observable universes.

This aspect of chaotic inflation was later generalized to include all versions of inflation. It is now believed that once inflation begins, it will never end, a concept called *eternal inflation*. The inflationary era was originally expected to end because the false vacuum is inherently unstable and decays, similar to a radioactive isotope. Like the decay of uranium or plutonium, the rate of decay of the false vacuum is mathematically exponential, but recall that the rate of expansion during the inflationary era is also exponential. It turns out that in most models the rate of inflationary expansion is significantly faster than the decay of the false vacuum, with the overall effect that the false vacuum actually gets bigger, not smaller, with time! Individual "pockets" of the false vacuum will decay at different times, each leading to a separate universe, dubbed *pocket universes*. The collection of all pocket universes that derive from the original false vacuum is sometimes called the *multiverse*. According to eternal inflation, at all moments, some pocket universe is being created by undergoing inflation, and once formed it evolves independently of all its sibling universes. For example, our pocket universe began about 13 billion years ago in an inflationary era we generically call our "big bang," and as far as we know our universe will continue to exist (albeit becoming a darker and lonelier place) for an infinite period of time. Does this mean that inflation has always occurred somewhere in our multiverse? Has a discussion of the ultimate origin of the universe really been avoided? It appears that, at least in simple inflation models, there is still a finite beginning somewhere in the distant past, and inflation is only eternal in the future direction.

No matter how "pretty" (or mindblowing) the theory, the scientific method demands that the predictions of any theory be tested to see if it is a consistent and useful description of the observable world. Among the testable predictions of most inflationary models are the follows:

1. $\Omega = 1$
2. The spectrum of density perturbations produced should be nearly (but not exactly) scale invariant. This means that although the density perturbations would come in different sizes, smaller ones

would contribute in nearly the same way as the large ones. This is the same form assumed in most models of structure formation in the early universe.

3. There should exist a distinctive background of gravity waves left over from the inflationary epoch.

In the early 1990s, inflation was left hanging in terms of observations. Ω was considered to be closer to 0.3 than 1, there was no precise measurement of the density perturbations, and gravity waves had not been detected. Yet many cosmologists felt confident that it was only a matter of time before observations caught up to theory.

DARK MATTER REDUX

The increasing evidence for dark matter has already been laid out for the reader, with the result that by the 1990s it appeared that the universe was dominated by dark matter over visible matter by a factor of 20:1. The thought that 95 percent of the universe is made of some invisible substance only detectable through its gravitational influence is certainly unsettling, and leads to an important question—what is the nature of this mysterious material?

A natural suggestion was that it was made of some small, dark nuggets of normal matter that emit little to no light, such as planets, brown dwarfs, and stellar corpses. These candidates for dark matter were lumped together under the term MACHOS, short for massive astrophysical compact halo objects. Several studies (with rather tongue-in-cheek names such as EROS, OGLE, and DUO) utilized microlensing to search for MACHOS. In microlensing, a small object like a planet or star passes directly in front of another small object, causing a distinctive brightening of the image through a gravitational focusing of the light. Unfortunately, although these studies did discover MACHOS, it was in disappointingly small numbers. It was clear that MACHOS could not contribute significantly to the dark matter.

There is also a theoretical reason for discounting MACHOS as the main contributor to dark matter. Normal (or baryonic) matter is comprised of neutrons and protons. The percent of deuterium generated during big bang nucleosynthesis is exceedingly sensitive to the density of baryonic matter. Observations of the current amount of deuterium in the universe (and other light isotopes created in big bang nucleosynthesis) limit the amount of baryonic matter in the universe to about $\Omega_B = 0.04$ to 0.05. This means that while a small amount of dark matter can be in the form of normal material made of protons and neutrons, the vast majority must be made of some exotic material.

Nonbaryonic dark matter candidates can be broken into two classes—*hot dark matter* (HDM) and *cold dark matter* (CDM). The former is called "hot" because its particles move very fast, near the speed of light, and the latter "cold" because they are more sluggish. The main candidate for HDM is the neutrino, which has such a small mass that it travels close to the speed of light. Unfortunately, because HDM travels so quickly it resists clumping together into structures unless there are a lot of particles. If HDM dominates the universe, it would tend to smooth out any clumping at small scales in the early universe, with the result that huge, flattened structures several hundred megaparsecs in diameter would form first in the universe, only to slowly fragment into smaller structures such as galaxy clusters and individual galaxies over time. This "top-down" or pancake model of structure formation was first proposed by Russian theorist Yakov Zeldovich in 1970.

Neutrinos have one very specific advantage as a dark matter candidate—they are definitely known to exist! The leading candidates for cold dark matter—*weakly interacting massive particles* (WIMPS)—do not enjoy this luxury. The favored candidates for WIMPS are the supersymmetric partners predicted by supersymmetry. For example, each neutralino, the SUSY partner of the neutrino, would have a mass 100–1000 times that of the proton. The problem is that SUSY is still a speculative theory and no SUSY partners have been discovered. Another candidate for WIMPS is the axion, the as-yet unseen particles predicted by theories that seek to explain how matter won out over antimatter.

Despite this rather sizeable (but not insurmountable) problem, cold dark matter is the reigning explanation for dark matter for one very important reason—it correctly predicts the structures seen in the universe today. Since CDM moves slowly (relatively speaking), it clumps together much more readily than HDM, predicting that smaller structures such as galaxies should form first, and later aggregate into clusters and larger structures. More specifically, galaxies form within dark matter haloes, condensing out of gas and later fragmenting into stars. The dark matter halo remains surrounding the galaxy, as in the case of our own Milky Way. The dark matter haloes of individual galaxies later gather together to form galaxy clusters, and so on, resulting in the pattern of filaments seen in redshift surveys. In addition, surveys of galaxies and galaxy clusters show that most galaxies were formed at redshifts $\sim$1 to 4, with the earliest formed within the first billion years of the universe ($z > 6.5$), while galaxy clusters formed much more recently, around $z \sim 1$

or less, and superclusters are still coming together, consistent with a CDM-dominated universe.

DARK ENERGY

The Accelerating Universe

Despite the successes of the CDM model of the universe, there were still serious problems in the late 1990s. Firstly, recall that the inflation predicted that $\Omega = 1$ (the universe is flat) yet even with dark matter the universe appeared to be open, with $\Omega \sim 0.3$. In addition, the ages of globular clusters seemed older than the age of the universe predicted from the Hubble constant, which was settling down at a value around 75 km/sec/Mpc. All these issues could be addressed if Einstein's long-maligned cosmological constant was not zero, but most cosmologists were averse to involving this "fudge factor" without significant observational evidence. Such evidence was discovered by two independent teams of astronomers in 1998.

Recall that the value of Hubble's constant is expected to change over the evolution of the universe, as the expansion of the universe changes speed. It was expected that the expansion of the universe should be slowing down, or decelerating, hence the variable that measures the rate of change of the expansion was dubbed the *deceleration parameter* (q). This parameter itself can change values as the universe ages. In order to measure q, cosmologists need to plot a Hubble's law diagram looking far into space (literally back in time) for redshifts greater than $z = 0.1$, where Cepheids are too distant to be seen and type Ia supernovae are the most trusted standard candles. What the two teams found was astounding—distant supernovae were fainter than they should be given their distance, meaning the universe had expanded faster, not slower, during the intervening time. In other words, in the recent past (and still today), the expansion of the universe is accelerating, not decelerating, and the value of q is therefore negative. More curious still, supernovae farther away ($z \sim 1.7$) looked brighter than would be expected for an accelerating universe, which meant that back that far in time the universe was still decelerating!

Stop, take a breath, and close your eyes for a moment, because the universe is about to get a whole lot weirder in the next sentence. The observations described in the previous paragraph seem best explained if the universe really is flat ($\Omega = 1$), but nearly 30 percent of the universe is composed of dark matter and most of the other 70 percent is made of some other invisible component which acts as a repulsive force, called *dark energy*. Based on the supernova observations, the expansion history of the universe can be divided into three broad epochs. During the first few hundred thousand years of its history, the evolution of the universe was dominated by the radiation (photons) it contained, and the expansion of the universe slowed down over time. After this, the evolution of the universe was matter-dominated until $z \sim 0.5$, corresponding to about 4 billion years ago. During the matter-dominated epoch the expansion rate of the universe continued to slow down, but at a slightly different rate than occurred in the radiation-dominated era. About 4 billion years ago, dark energy began to dominate the evolution of the universe, and the expansion rate began to accelerate. The reason for this is that unlike radiation and matter, whose density thins out as the universe expands, dark energy is a quality of space itself, so its density stays the same. At some point, the universe had expanded so much that the density of the dark energy was finally greater than the density of matter and radiation, and its repulsive nature began to dominate the evolution of the universe. Because of this accelerated expansion, the universe is actually older than one would predict from the current value of Hubble's constant, and there is no longer a conflict between the age of the universe and the ages of globular clusters. This is reminiscent of Georges Lemaître's use of a cosmological constant to try and relate the high values of H_0 derived in the 1920s and 1930s to the accepted age of the Earth.

Several questions immediately come to mind. Firstly, how reliable are these supernova measurements? Supernovae seen at large distances are seen as they were much earlier in the universe. Are supernovae created by previous generations of stars the same as supernovae that occurred more recently? Is interstellar or intergalactic dust skewing the brightness measurements? While astronomers investigate these important points, alternate ways of affirming the acceleration of the universe continue to be explored.

Candidates for Dark Energy

Perhaps more immediate is the problem of figuring out the nature of this dark energy. Many cosmologists simply use the terms dark energy and cosmological constant interchangeably, and even use the Greek letter lambda (Λ), which was used in equations to denote the cosmological constant, in describing dark energy. But care needs to be taken. The cosmological constant is generally pictured in a very specific way—as the vacuum energy (the inherent energy density of the vacuum itself). Recall that the vacuum is not really empty, but is instead a seething sea of virtual particle-antiparticle pairs being created and annihilated due to quantum effects. However, particle physics currently predicts that when adding up all the contributions of all possible particle-antiparticle pairs, the energy density of the vacuum should be more than 10^{120} times greater than the energy contained in all the normal matter in the universe! If the cosmological constant/dark energy were really that large, it would rip apart all matter in the universe, down to the level of protons and neutrons. The fact that the predicted and observed limits of the vacuum energy density differ by such a catastrophic amount is sometimes called the *cosmological constant problem.* As theorist Leonard Susskind of Stanford University quipped,

Being wrong by one order of magnitude is bad; two orders a disaster; three, a disgrace. Well the best efforts of the best physicists, using our best theories, predict Einstein's cosmological constant incorrectly to 120 orders of magnitude! That's so bad that it's funny. (2006, 66)

If dark energy isn't the vacuum energy, then what is it? One popular suggestion is called *quintessence.* This is a proposed new type of particle or field whose properties change in time as the universe evolves. Quintessence acts like the inflaton field, in that it has a negative pressure and is repulsive, accelerating the expansion of the universe. It is much weaker than the inflaton field, so it does not cause as great an acceleration as during inflation. The fact that the quintessence field varies also differentiates it from being just another version of inflation. Since quintessence changes, it is possible that if it were the correct explanation for dark energy that the acceleration currently observed could cease in the future. A third contender is actually a unified explanation for dark energy and dark matter called quartessence, based on theories of exotic fluids. Finally, the strangest candidate is called phantom energy, an extremely unstable quantum field that would cause so much acceleration that it could cause the Big Rip—the shredding of all structures down to the atomic level. Some comfort can be gained from the fact that if such a runaway dark energy does exist, we would not have to fear the evisceration of the universe for another 20 or so billion years. The reader should take from this discussion the idea that dark energy is an active playground for theorists, and until observations can definitively discriminate between the predictions of these different candidates, the theoretical gymnastics will continue.

Putting aside the rather obvious demise predicted by the Big Rip, how does the existence of dark energy change the future evolution of the universe from that described previously? The key difference is that a continued accelerated expansion will create a horizon around every observer in the universe, a distance beyond which one cannot observe anything because light from those objects has not had time to reach him or her. One by one galaxies will disappear behind this impenetrable curtain. Computer models show that if the universe is dominated by a cosmological constant, by the time it is about 100 billion years old, the only galaxies we will be able to observe from Earth are those of the Local Group, or more accurately, whatever exists in our neighborhood after the merger of the Milky Way and M31.

THE CONCORDANCE MODEL

Inflation, cold dark matter, dark energy—three extraordinary concepts which combined have given cosmologists the current best description of the universe we inhabit. Called the *inflationary Λ CDM model*, it is based on a number of measurable parameters, including the Hubble constant, the deceleration parameter, the age of the universe, the temperature of the cosmic microwave background, and the densities of baryonic matter, dark matter, neutrinos, and dark energy. As with any scientific model, it must be vigorously tested, and as we shall see, it certainly has. Because of its success in meeting every major observational challenge thrown its way, the inflationary ΛCDM model has been dubbed the concordance model. At this moment, its weakest link is actually the assumption that dark energy is described by a cosmological constant modeled on the vacuum energy. Much of the observational support for the concordance model has come from exquisite details in the cosmic microwave background, caused by sound waves in the early universe, a curious modern twist on the ancient idea of the music or harmony of the spheres.

Acoustic Waves and the Early Universe

Aristotle wrote in his *Metaphysics* how the Pythagoreans ascribed numerical values to the musical scales, which could then be related to the universe at large, especially the motions of the planets. This "music of the spheres" could not be heard by any but the especially gifted (reportedly including Pythagoras himself). Aristotle himself believed the idea to be beautiful but illogical. Ptolemy, as well, wrote in the third book of his *Harmonica* that the motions of the stars matched up with musical scales. Boethius, the Roman philosopher, wrote about three types of music: that of the universe, human music, and music created by instruments.

He believed that the first was the most important to study in order to understand the motions of the heavens.

Johannes Kepler gave the idea serious consideration (one might say obsessively so), in his *Harmonices Mundi*. He discovered that the ratio of certain properties of planetary and lunar motions were approximately the same numerical value as that between the notes in chords. On the basis of the observational relationships he found, he constructed musical scales composed of notes representing each of the planets. Interestingly, in order to do so, he had to ignore the octave of the notes (e.g., lowering the "sound" of Mercury by six octaves in order to fit it into the scale). Despite the idea's poetic qualities, the concept of the music of the spheres eventually became little more than a historical footnote in the minds of scientists until modern day. With the rise of the concordance model, the existence of sounds generated in the early universe became an important means of testing the predictions of modern cosmology.

Sound is created by the propagation of pressure waves through a medium, such as air. Since the word "sound" is usually thought of in connection with the sense of hearing, a more generic term is acoustic wave, which encompasses frequencies both too high and too low for the human ear to register. As the trailers for the film *Alien* correctly warned, "in space, no one can hear you scream." In general, interstellar space is currently so close to a perfect vacuum that sound waves cannot propagate. However, in astronomical locations where the gas density is high enough, acoustic waves can and do occur. For example, sound waves can travel through the atmospheres of Venus, Mars, and Saturn's moon Titan, as well as Earth. Stars, as hot spheres of dense gas, are also laboratories for the detection of acoustic waves. In 1962 it was discovered that the visible surface of the sun—its photosphere—is bubbling with a period of about 5 min. This "five minute oscillation" was found to be a global rather than local phenomenon caused by sound waves (called p-modes) traveling through the sun. Helioseismologists have monitored these oscillations for a number of years through the Global Oscillation Network Group of ground-based telescopes (GONG) as well as the Solar and Heliospheric Observatory (SOHO). Such studies allow astronomers to map the sun's interior.

But how could acoustic waves form in the vacuum of space, and what do they have to do with radiation? The key point is that although interstellar space is unbelievably cold and relatively empty at this point in the universe's evolution, during the first few hundred thousand years or so it was nothing of the sort. In contrast to the popular misconception fostered by the term "big bang," the universe began in silence, because there were no initial acoustic waves. However, inherent inhomogeneities in the density of matter and energy are predicted by inflationary models. These small density perturbations created a series of gravitational "valleys." Light and matter (bound together in the opaque fog of the universe before recombination) fell into the valleys and compressed. The compressed radiation exerted pressure and the matter and radiation shot out of the valley, only to settle into another valley. Dark matter, which does not interact with matter and radiation, also settled into the valleys, gradually making them bigger by its gravity. Because matter was compressed and expanded, it generated pressure waves, and because the radiation was coupled to the matter, the behavior of the radiation was affected by these pressure waves.

Results from the Wilkinson Microwave Anisotropy Probe (WMAP)

Our story now returns for the moment to more familiar ground, the cosmic microwave background. Detailed measurements of the CMB were made by the Cosmic Background Explorer (COBE) in 1992, which verified the effective temperature of the radiation (2.725 K) and discovered slight variations in temperature across the sky to about one part in 100,000, a reflection of slight density perturbations in the early universe, like those predicted by inflation. It also found that over angular scales of 10–90 degrees there was nearly equal power on each scale, agreeing with the nearly scale-invariant spectrum predicted by inflation. Scientists who supported inflation were understandably ecstatic. George Smoot, the lead investigator of the project, told journalists, "if you're religious, it's like seeing the face of God." Stephen Hawking called it "the discovery of the century, if not of all time" (Singh 2004, 462–463). Not surprisingly, COBE scientists Smoot and John Mather were awarded the 2006 Nobel Prize in Physics for these discoveries. Angular variations or anisotropies smaller than about 7 degrees unfortunately could not be seen because of the satellite's limited resolution. It was these smaller angular anisotropies which would later provide some of the strongest evidence for inflation and the concordance model.

It has become standard procedure to describe variations in the CMB in terms of mathematical functions called spherical harmonics. An analogy can be found (in a single dimension) in a guitar string. The fundamental tone of the string is such that the entire string moves back and forth except the two points where the string is connected to the guitar. The first harmonic tone occurs when there is one stationary point on the string other than the ends, and so forth. In terms of spherical harmonics,

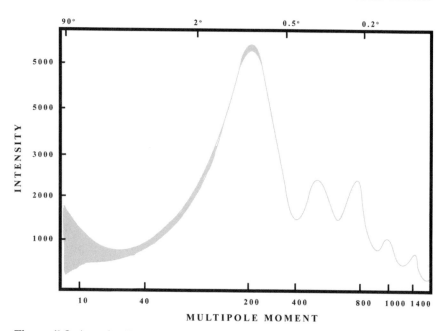

Figure 5.2 Angular Power Spectrum of the Cosmic Microwave Background

these are described by the angular frequency l, where l = 0 represents the fundamental tone, the first harmonic, l = 1, etc. In terms of the CMB, cosmologists plot the intensity of temperature variations versus l, creating a graph, as seen in Figure 5.2, called the *angular power spectrum*. Since l = 180/θ, the larger the value of l (or multipole, as it is called), the smaller the angular size of the variations. The most powerful signal in the CMB is the dipole radiation (l = 1, θ = 180°). In one direction of the sky the CMB is 3.36 mK (thousandths of a Kelvin) warmer than in the other, due to the fact that the solar system is moving in this direction at 370 km/sec. This signal is more than 100 times stronger than other angular anisotropies and does not give us information about the early universe, so it is subtracted out before the angular power spectrum is graphed. Since COBE could not see angles less than 7 degrees, it could not pick up multipoles larger than about 25 or 26. The ability to see higher multipoles is vitally important, as theory predicts acoustic waves in the early universe led to several pronounced peaks at multipoles higher than l = 50 (or angles smaller than a few degrees).

The sizes of the primordial acoustic pressure waves varied, from the longest or "fundamental" to smaller "harmonics," and depended on the speed the waves traveled and how much time had elapsed from the start of the universe. Since the speed of sound in this dense "soup" was approximately 60 percent the speed of light, in the estimated 400,000 year span of this acoustic era the longest waves (and deepest "tones") that could be generated were approximately 220,000 light years in length. This should appear in the CMB as a noticeable peak at around 1 degree (or $l \sim 220$). Smaller peaks should also appear at multiples of this multipole (or fractions of a degree). These are collectively known as the acoustic peaks. At recombination when electrons formed stable atoms and radiation ran free, the speed of sound slowed dramatically, and eventually stopped, and the waves "froze" into place as the pressure ceased. Matter then fell into the gravitational wells, which had been enlarged by dark matter, and entered the road to the formation of astronomical structures. Multipoles over $l \sim 1000$ are smeared out and blend together into a smooth curve. The acoustic waves responsible for these multipoles are so small that their effects are overrun by interactions between photons and electrons. This is called Silk damping.

Thanks to the Wilkinson Microwave Anisotropy Probe (WMAP), launched in 2001, the existence of the first three acoustic peaks has been confirmed and the angular location and relative strengths of the first two peaks are precisely known. This is vitally important, because the exact angular location of the acoustic peaks as well as the relative heights of the peaks depend very dramatically on cosmological parameters such as the densities of matter and radiation and the geometry of the universe. The details of the WMAP angular power spectrum are not only strong evidence for the concordance model, but help to further hone the values of some of its important parameters. For example, the WMAP data fits exceedingly well with a flat ($\Omega = 1$) universe where normal matter made of baryons contributes 4 percent, dark matter contributes 26 percent, and dark energy makes up the majority of the density of the universe, at 70 percent. The exact identity of dark energy cannot be confirmed based solely on the WMAP observations, but various models under consideration are somewhat constrained by the observed data. It also supports a value of Hubble's constant around 73 km/sec/Mpc, in agreement with the Hubble Space Telescope studies. The best estimate for the age of the universe in 13.7 billion years, in agreement with the most recent studies of the ages of globular clusters. WMAP gives additional support for inflation (although it has not reached the sensitivity necessary to distinguish between the predictions of the various models of inflation). Recall that in general inflation predicts a nearly scale invariant spectrum of density perturbations, which should be reflected in the cosmic microwave background. WMAP has confirmed that this is the spectrum observed, providing some of the strongest evidence yet for inflation.

Not only have the WMAP observations provided evidence for the inflationary ΛCDM model, but they have also largely ruled out other competing models. For example, Israeli physicist Mordehai Milgrom proposed an alternative to dark matter in the 1980s. Known as Modified Newtonian Dynamics (MOND), the theory postulates that, on the scales of galaxies, nature does not exactly follow Newton's law of gravity, with these modifications giving rise to observations such as the flat rotation curves of spiral galaxies that most cosmologists attribute to dark matter. Not only does MOND fail to account for observations supporting dark matter in galaxy clusters and has difficulty reproducing the currently observed large-scale structure, but it also predicts a "much lower third [acoustic] peak than is observed by WMAP" (Spergel, et al. 2006, 14).

Other Observations

So far we have focused on the relic anisotropies in the cosmic background radiation caused by the acoustic waves generated by matter before recombination. When recombination occurred and the radiation decoupled, clumps of matter were expanding and contracting independently and were frozen in various stages of this pattern. This is best understood by tossing a pebble into a pond, and after the ripples have expanded some distance, freezing them into place. There should therefore be similar acoustic peaks in the distribution of galaxies. This was verified by the Sloan Digital Sky Survey (SDSS) in 2005. In agreement with the predictions of the inflationary ΛCDM model, there is a 1 percent overdensity of galaxies with separations of 150 Mpc. The SDSS and other large-scale galaxy surveys have also verified two other cosmological predictions involving observations of distant galaxies. The first, called *cosmic shear*, is a weak kind of gravitational lensing. Fluctuations in the distribution of dark matter throughout the universe distort the shapes of distant galaxies but do not cause enough lensing to create multiple images, as in so-called strong gravitational lensing. Observations of cosmic shear support the predictions of the inflationary ΛCDM model. The second observation, related to cosmic shear, is called *cosmic magnification*. Light from distant quasars is focused by intervening dark matter in such a way as to make the quasars appear slightly brighter than they should otherwise. The magnification of 200,000 quasars was discovered by SDSS in 2005. The results were also in agreement with the concordance model.

Has cosmology been solved? Is the concordance model the ultimate description of the universe? The scientific community has certainly learned a valuable lesson from the rather conceited comments of Lord Kelvin, who in 1900 declared that "There is nothing new to be discovered in physics now. All that remains is more and more precise measurement" (Davies and Brown 1988, 3–4). It is generally accepted that the predictions of the inflationary ΛCDM model have been quite successful at fitting the observed universe, but there are still observations to be conducted with improved technology. For example, the direct detection of gravity waves will provide a serious test of the model. It should also be noted that the WMAP observations taken in isolation do not demand that dark energy exists, but rather confirm that the data fits very well (some might argue best fits) with a flat universe composed of 70 percent dark energy—whatever form it might take. Further observations of the CMB and distant quasars and galaxies should eventually be able to differentiate between different versions of inflation as well as the different explanations for dark energy. So even if the inflationary ΛCDM model emerges as the undisputable champion of cosmological theories, there is enough work to do to keep researchers and graduate students busy for years to come.

But what of dissenting voices, and others who constantly push the boundaries of current knowledge? In science, multiple models emerge at any given time and are subjected to scrutiny and peer review. No model is considered final, but always opens the door for further avenues of exploration. Our journey now turns to these speculations.

6

THE ROAD GOES EVER ON: CONTINUING SPECULATIONS

MULTIPLE DIMENSIONS

According to Einstein's general theory of relativity, space and time are interwoven into the flexible four-dimensional fabric of space-time. But could there be more dimensions of space that we do not experience in everyday life? In the 1921, Polish mathematician Theodor Kaluza utilized an extra dimension of space to try and unify gravity and electromagnetism. In 1926 Swedish mathematician Oskar Klein demonstrated that this extra dimension was unobservable because it was rolled up, or compactified, into a circle the size of the Planck length (10^{-35} m). Unfortunately their unification scheme did not succeed, but it opened the door for further investigations of the physics of higher dimensions. Princeton theorist Edward Witten warns that "the idea of extra dimensions might sound a little bit strange to anyone who hasn't studied physics. Anyone who has gone into physics professionally, will know that there are many things that are a lot stranger than extra dimensions" (Davies and Brown 1988, 101). We will now put his observations to the test.

String Theory

In 1968, Italian physicist Gabrielle Veneziano proposed a theory that involved extra dimensions as an explanation for the every-growing number of hadrons (before the acceptance of the quark model). His model became known as *string theory*, and visualized the interactions of the strong force in terms of bosonic strings. In 1970 John Schwarz and André Neveu developed a fermionic string theory, opening the door for further study. With the success of QCD, string theory was no longer "necessary" to describe the strong force, and seemed doomed to the dustbin of the scientific method. John Schwarz and Joel Scherk rescued it in 1974, recognizing that string theory was actually a theory that included gravity, leading to its revision as a candidate for a unified theory of particles. Each unique frequency of vibration of a tiny bosonic or fermion string (each only about 10–35 m long) was postulated to be what we observe as a different elementary particle. But since bosons mediate forces and fermions make up matter, having two separate theories seemed contradictory to a successful model of unification.

String theory fell out of favor until 1984, when Michael Green and John Schwarz combined string theory with supersymmetry, creating a theory that unified bosonic strings and fermionic strings into a theory of *superstrings*. Superstrings could be closed (have their ends join together to form a tiny loop) or open (the ends remain unattached). Closed strings appeared to give rise to gravitons, the supposed particle which mediates gravity in models that unify general relativity and quantum mechanics. The price to be paid for this seeming unification was that superstring theory was mathematically consistent in ten dimensions, while the real universe seems to exist only in four (three of space and one of time). Following the work of Kaluza and Klein, the "extra" dimensions were said to be curled up into tiny little knots far too small to be observed, called Calabi-Yau manifolds.

Superstrings soon became the new fashion in theoretical physics, but not without making vocal enemies. Proponents hailed its elegant mathematics, but beauty alone is not a reason for accepting a highly speculative theory. Indeed, Ralph Alpher and Robert Herman of big bang fame warn that

Science brings some problems upon itself. For example, scientists are frequently attracted by a theory because they find it beautiful, when in fact it may be wrong.... All this reflects the fact that the doing of science is a human and creative process. (2001, 173)

Leading theoretical physicist Stephen Hawking cut to the chase and declared string theory "pretty pathetic" for its lack of ability to make predictions observable with current technology (Hawking and Penrose 1996, 123). Others charged that superstring researchers were receiving an unfairly large amount of available grant money (much of it received from the U.S. Departments of Defense and Energy) and were snatching up some of the best graduate students for research that was speculative at best. String theorist Gary Horowitz noted that although many of the remarks by colleagues have been "disparaging, the large number of them at least seems to indicate that string theory is quite important!" (Hawking and Penrose, 1996, 135).

Despite the strong opinions of physicists on both sides of the string controversy, in the end only the scientific method truly matters. But string theorists have openly admitted that superstring theory does not neatly fit into the "hypothesis-predict-test" model of science. MIT's Lisa Randall explains that

An enormous theoretical gulf separates string theory, as it is currently understood, from predictions that describe our world. String theory's equations describe objects that are so incredibly tiny and possess such extraordinarily high energy that any detectors we could imagine making with conceivable technologies would be unlikely ever to see them. Not only is it mathematically tremendously challenging to derive string theory's consequences and predictions, it is not even always clear how to organize string theory's ingredients and determine which mathematical problem to solve. (2005, 69)

It is therefore possible that string theory may have little to do with our observed universe, and it would be very difficult to know one way or the other. Despite the serious challenges in directly testing superstrings, there are several predictions hopefully within the grasp of reasonable technology to test. Superstring theory predicts a specific spectrum of gravity waves, which may be verified or refuted once gravity waves are directly detected. Superstring theory also predicts small imprints on the cosmic microwave background that may be detectable with improved technology. On the negative side, if the concordance model of cosmology continues to gain further observational support, string theory will be thrown into question, because superstrings are generally not thought to be consistent with the existence of a cosmological constant.

M-theory and Braneworlds

For several reasons, superstring theory was once more "out of fashion" in the early 1990s, not the least of which being that there were five competing and distinct models. If superstrings were purporting to unify the forces of nature, shouldn't they first unify themselves? A possible answer soon came in the form of the holographic principle, discovered by Dutch physicist Gerard 't Hooft. This principle asserts that for some region of space, all the information contained inside it can be represented by the region's boundary. This is similar to a hologram, a two-dimensional "projection" of a three-dimensional object made with lasers. One of the most important special cases of the holographic principle is the AdS/CFT correspondence or duality. Also named the Maldacena conjecture, after its Argentine creator, Juan Maldacena, the acronym stands for *Anti-de Sitter/ Conformal Field Theory.*

Anti-de Sitter space is a variation of the de Sitter space used in inflationary models, and has a negative cosmological constant. It has a relatively simple geometry, which means that its properties have been extensively studied, and even though it is not an especially realistic model of the observed universe, calculations done using this space are more simple than in other geometries. A conformal field theory is a theory of fields in which the equations have a particular mathematical symmetry (i.e., the theory does not vary under certain mathematical transformations, such as multiplying all the factors by the same constant or rotating them by the same amount). A *duality* connects two apparently different physical theories and shows that they are equivalent. This is extremely useful when one theory is easier to calculate than the other. The most widely known example in physics is electromagnetism. The electric and magnetic fields (and the corresponding charges) can be interchanged in Maxwell's equations and the equations will remain the same.

Using these mathematical principles, Princeton string theorist Edward Witten demonstrated in 1995 that the five superstring theories as well as supergravity theory itself were nothing more than special cases of a much more fundamental, underlying theory in eleven dimensions which he dubbed M-theory. The reason for the name of the theory, as well as its precise characteristics, are not well understood. The M is alternately said to stand for "magic, mystery, or membrane, according to taste" although "critics might prefer 'mythical' or 'mystical'" (Naeye 2003, 40). Others have suggested it stands for "mother of all theories." The key is that M-theory predicts that besides point-like (zero-dimensional) particles and one-dimensional strings, there are two-dimensional membranes. There are also higher dimensional objects generically called *branes*.

The AdS/CFT correspondence states that a unified theory of quantum gravity (e.g., a superstring or supergravity theory) defined in an anti-de Sitter space-time of four spatial dimensions is equivalent to another quantum field theory living on the three spatial dimensional boundary of the AdS space. In various "braneworlds," the boundaries are depicted as branes (one of which represents the universe we experience) and the intervening space of four spatial dimensions is called the bulk. Other higher dimensions are either compactified or otherwise "distracted" in such a way as to not be directly obvious to us. In some sense, our universe is merely a shadow being projected by the bulk dimension. Our universe is the boundary and the bulk is the bounded region. Therefore, studying physics on the boundary (in our universe) tells us about the physics of the higher dimension, and vice versa. It is as if we are shadow puppets, and by studying the details of the shadows we can figure out the way the hands are arranged in higher dimensions.

One of the most important braneworld models was developed by Lisa Randall and Raman Sundrum in 1999, and was used in an attempt to tackle one of the fundamental lingering roadblocks in the path of unifying the forces, namely why gravity is so much weaker than the other forces. Most people have the mistaken impression that gravity is strong, especially when trying to climb several flights of stairs. But the fact that it is done so routinely clearly shows the weakness of gravity. The simple action of hanging a note on the refrigerator using a magnet also demonstrates gravity's inherently wimpy nature. The electromagnetic force that holds our atoms together is 10^{43} times more powerful than gravity. This relative weakness of gravity is called the *hierarchy problem*, and has no obvious solution in the standard model of particle physics.

The key is the way the different forces are described in string theory. Recall that closed strings represent gravity (through the graviton), while open strings represent the particles of matter as well as the particles that mediate the other three forces. The open ends of these strings attach to the brane that represents our universe and prevents these particles or their corresponding forces from leaking into the bulk. The closed strings are not confined to our brane and leak into the bulk, and can actually travel to any other branes contained in that particular model. The rate of leaking can be carefully balanced such that it solves the hierarchy problem yet is not so drastic that *current* experiments would notice the difference in gravity's behavior. It is possible that future particle accelerators with much higher energies may be able to find small variations caused by gravity's proposed leaky behavior.

Braneworlds have also been used as alternatives to the inflationary cosmology. In 2001 Paul Steinhardt, one of the fathers of "new inflation," and colleagues created what they felt was a direct competitor to the earlier theory. In their model, our universe is represented as a "visible brane" which is located some finite distance in the bulk away from a "hidden brane." The universe was initially empty and cold, which changed when the two branes collided, bouncing or passing through each other. The collision converts some of the kinetic energy of the hidden brane into particles and radiation on our brane, and from there the model evolves similarly to the hot big bang. This theory was dubbed the *ekpyrotic model,* named after the Greek word for conflagration, a nod to the model of the ancient Greek school of the Stoics who claimed the universe was born from fire. One selling point of the theory is that it avoids an original singularity, as the laws of physics do not break down at any point in the process.

A year later, two of the ekpyrotic universe's creators, Steinhardt and Cambridge physicist Neil Turok, expanded the idea to encompass a cyclic model of fiery collisions and expansions which includes a quintessence model of dark energy as a central component. In this cyclic model, the general homogeneity and isotropy of the universe as well as the density perturbations are explained as relics of the transition between cycles. Both the cyclic model and inflation predict density perturbations that are so similar that current technology cannot distinguish between the two models. When gravity waves are finally detected, cosmologists will be able to compare the relative success of the two models more carefully, as inflation predicts distinctive gravity waves while the cyclic model predicts no primordial gravity waves.

COSMIC STRINGS

Inflation and its ekpyrotic rival are not the only theories proposed to explain the primordial density perturbations that seeded structure in the universe. In grand unified theories, topological defects or "snags" would form in the fabric of space-time as various symmetries were broken in an imperfect way. These would be areas of high energy density, very massive, and come in a variety of dimensions. For example, point-like defects are called monopoles (such as the magnetic monopoles previous discussed), one-dimensional defects are called cosmic strings (to prevent confusion with string theory) and two-dimensional defects are termed domain walls. In the seminal paper on the subject in 1976, T.W.B. Kibble demonstrated that the existence of domain walls can be ruled out because of their gravitational effects. They have an inherent repulsive gravitational field and produce large-scale structures different from what is observed in the universe today. While theorists Kolb and Turner called domain walls "cosmological bad news," they acknowledged that cosmic strings are "much more palatable to a cosmologist" (1990, 220).

Cosmic strings would have less drastic effects on the universe, and it was suggested that they could act as the density perturbations or seeds for the formation of galaxies, a problem which was actively being researched in the 1980s. Because of this, a flurry of theoretical papers soon appeared describing both the properties of cosmic strings and possible observational tests of their existence. Cosmic strings were theorized to come in two varieties: infinitely long, open strings, and loops of closed strings. The latter were envisioned to have been created when open strings intersected and chunks broke off and closed. Both versions were pictured as very thin (approximately 10^{-30} cm) and have an unbelievably high density—10^{15} tons per inch! Cosmic strings of both types were envisioned as possible sources of structure in the universe. The closed loops were thought to be able to attract matter to them as they oscillated close to the speed of light, while the open strings were pictured as moving through the universe, creating a gravitational wake in the process. Matter would tend to fall into these wakes and form structures. After creating these structures, closed strings would dissipate through the emission of gravity waves, and today all that might remain would be the structures they seeded, as well as relic gravity waves.

With improved observations, the likelihood of cosmic strings even existing, let alone taking an important role in structure formation, has diminished significantly. Cosmic strings make two important predictions for structure in the universe—that it would be strongly skewed to line-like features, and that the density perturbations would be of a different form than those predicted by inflation. Neither of these predictions appear to match the universe we inhabit. Additionally, the gravitational fields of open strings would produce very distinctive patterns of gravitational lensing, which have not been observed. Finally, as an open string moved, it would create distinctive discontinuities in the cosmic microwave background, called the Kaiser–Stebbins effect. This has also not been observed. It is therefore clear that if cosmic strings exist (and there is no definitive evidence in their favor), they are ruled out as a significant source of structure formation in the early universe. As with all the scientific speculations discussed so far, cosmic strings were "voted off the island" not as a result of their personality but because they failed to meet the rigorous tests demanded by the scientific method. Some theories (such as superstrings) are harder to test, but in the end, all theories must submit to the demands of the scientific method, even those that ask perhaps the most tantalizing questions of all—what is the likelihood of life elsewhere in the universe?

THE FINE-TUNED UNIVERSE

In 1938, famed physicist Paul Dirac noted a curious relationship between the numerical size of various parameters in the universe, including its then-presumed age, namely that various combinations of these constants had values around 10^{39}, which became known as the large number hypothesis. Dirac believed that this could not be a coincidence, and instead revealed some important relationship between cosmology and the subatomic realm. Robert Dicke pointed out in 1961 that since the age of the universe would obviously change over time, the various other parameters would also have to change to keep their ratio close to Dirac's special value. Since this is not accepted by most physicists, he found another way to explain the supposed relationship. He pointed out the fact that we as observers can only exist for a certain period of the history of the universe. For example, we could not make observations of the universe before sufficient generations of stars had previously died and made enough carbon to construct our bodies. We can also only exist as long as there are stars and planets of just the right properties to sustain us.

This was the beginning of a realization that the universe seems to be "fine-tuned" in such a way as to make carbon-based life-forms possible. For example, if the strong force were much stronger, all protons would pair off and there would be no normal hydrogen (and hence no water). If it were much weaker, large atomic nuclei could not exist. If gravity had been much weaker, structures such as stars and galaxies would never have been created, but if it were much stronger, the universe would have recollapsed before now. If the universe had been created with more than the three spatial dimensions we currently enjoy, neither atoms nor planetary orbits would be stable, but in fewer than three spatial dimensions basic biological processes such as circulation of blood and digestion are not possible. Various explanations for this fine-tuning have been proposed by scientists, philosophers, and religious writers. Some have taken a nonscientific approach and explained this as evidence of some grand designer. Within the scientific sphere, the anthropic principle and multiverse theories are both somewhat controversial alternative explanations to the religious concept of fine-tuning by design.

The Anthropic Principle

A scientific discussion of the anthropic principle is best done within the framework of careful definitions, as there are several concepts housed under the "anthropic" umbrella. The most common version of the principle is the so-called "weak" version attributed to Robert Dicke that was discussed above. Essentially the cause of the fine-tuning of the universe is not explained, other than to say that if the constants were not the values they currently are, we as observers would not be here to observe them. In some ways it explains why intelligent life exists at this point in the universe rather than earlier, in that the conditions (such as the amount of carbon made within stars) necessary for life simply were not met until this point. Some have called this *weak anthropic principle* (WAP) a tautology with no predictive power.

The *strong anthropic principle* (SAP) was developed by Brandon Carter in 1974 and suggests that the universe had no choice in its fine-tuning because the eventual existence of intelligent observers was a necessity. Rather than to merely note that the fine-tuning of the universe makes our existence possible, the SAP seeks to answer "why" the universe is fine-tuned by appealing to our very existence. As speculative (and controversial) as the SAP is considered, even more so are the *final anthropic principle* (FAP) and *participatory anthropic principle* (PAP). The former claims that not only is the eventual existence of intelligent life a necessity for the universe, but that once it has arisen it will continue to exist indefinitely. The PAP is related to questions of the role of the observer in quantum mechanics that are beyond the scope of this book. In a nutshell, it claims that observers are necessary to bring the universe into full existence. In this view, events can only be said to have reality if they are observed by a human mind. However, the SAP, FAP, and PAP appear unlikely, as the only form of intelligent observers currently known to exist (humans) did not arise until the universe was well over 13 billion years old and well after all its basic structures (such as stars, galaxies, and superclusters) had already been formed.

Among the physics community, reaction to the anthropic principle(s) includes open hostility and resigned acceptance ("until there's a better explanation, it will have to do"). In their review of the standard model of the early universe, Kolb and Turner editorialized "It is unclear to one of the authors how a concept as lame as the 'anthropic idea' was ever elevated to the status of a principle" (1990, 269). String theorist Leonard Susskind also paints a rather vivid picture: "The Anthropic Principle affects most theoretical physicists the same way that a truckload of tourists in the African bush affects an angry bull elephant" (2006, 172). The anthropic principle even became the subject of parody; for example, mathematician David Shotwell coined the "entomologic principle" (the universe is fine-tuned to produce insects) and Carl Sagan offered a similar "lithic principle" (the universe is fine-tuned to produce rocks). On the other side of the spectrum, the anthropic principle has been used for many years by Stephen Hawking and other theoretical physicists to constrain the parameters of various models by selecting out those versions which allow for the existence of observers. For example, if a model allows a variety of types of universes, the anthropic principle can be used to only accept those types in which life is possible (because any types of universes that do not allow life are obviously not relevant to a discussion of our particular universe).

Multiverses

The other major scientific explanation for the fine-tuning of the universe is the multiple universe or multiverse paradigm. This is a set of theories which derives from the inflationary model of cosmology, based on the prediction that inflation is eternal. Separate inflationary regimes or "pocket universes" will continuously be produced as they nucleate out from the original false vacuum state. The collection of all these pocket universes (of which our observed universe is just one) is referred to as a *multiverse*. A possible explanation for the observed fine-tuning of constants in our pocket universe can be made on statistical grounds. Given a large (possibly infinite) sample of pocket universes, there should be a large range of possible combinations manifested for the fundamental constants. In at least one of those pocket universes the combination of constants is appropriate for the formation of intelligent carbon-based life, and this is the pocket universe in which we live. Other pocket universes could have conditions completely averse to the origin of life. There is no intention, no design, in some sense there is just dumb luck. Gordon Kane compares it to winning the lottery: "someone had to win, and no one selected who that was, except randomly. Just because a universe has a unique set of laws and parameters should not lead one to wonder whether that set was designed" (2002, 24). Again, the argument is that scientifically there is no need for a designer. Any theological discussions are separate and distinct, and are not ruled out by any of the scientific arguments presented here.

It has been suggested that the anthropic principle could be combined with the multiverse model, to provide an explanation for the tiny size of the cosmological constant. If it is an "environmental" variable and varies from one pocket universe to another, we can use the anthropic principle to state that we could only exist in (and hence observe) a pocket universe in which the value of the cosmological constant is small enough to allow for the formation of atoms, stars, and galaxies. In the words of cosmologist Michael Turner,

If the multiverse is right, it could be that we live in a little oasis. . . . Martin Rees has speculated that we could well be the only intelligent beings in the observable universe . . . We'd better be really careful not to extinguish it. (Yulsman 2003, 346)

THE SEARCH FOR EXTRATERRESTRIAL INTELLIGENCE

Are we truly the only intelligent species in the universe capable of cosmological queries? In 1600, controversial Italian philosopher Giordano Bruno was burned at the stake by the Inquisition. Among his crimes was proposing that "There are then innumerable suns, and an infinite number of earths revolve around those suns," and that on those Earths there could exist life (Munitz 1957, 183). In 1995, Swiss astronomers Michel Mayor and Didier Queloz discovered a strange planet orbiting the star 51 Pegasi with a period of 4.2 days. This was the first extrasolar planet, or exoplanet, discovered by the telltale gravitational wobble it inflicts as it orbits its parent star. The wobble manifests itself in a tiny, cyclical alternating blueshift and redshift of the parent star. Over 200 exoplanets have been discovered since this time, the vast majority by this Doppler oscillation method. It is sensitive to planets of large masses and small orbits, as the size of the spectral Doppler shift depends on the strength of the gravitational tug the planet inflicts on its star.

To find planets closer to Earth's mass, two other methods have employed—microlensing and transits. Microlensing can be used when a small object passes in front of another small object and the image of the background object is temporarily brightened through gravitational lensing. In the case of exoplanet searches, microlensing is well suited to finding planets of small mass, like the earth, and with larger orbits than those found through the Doppler shift method. In transit events, the planet crosses the face of the star, resulting in a characteristic dimming of the star. At least ten stars are known to have planets that can be monitored in this way. With the explosion of known exoplanets in the past decade, it is natural to speculate whether any of them (or the host of planets undoubtedly still undiscovered in our galactic neck of the woods) are currently inhabited.

The possibility of life elsewhere in the cosmos has fascinated scientists and laypersons alike for centuries. There is currently no scientific evidence that there is life elsewhere in the universe, despite uncorroborated reports of abductions by "little green men" and sightings of weather balloons, flocks of birds, and even the planet Venus incorrectly attributed to alien spacecraft. The scientific study of the possible properties of lifeforms that might exist outside of the earth is termed astrobiology, while the search for intelligent lifeforms outside of the earth is called *SETI*—the search for extraterrestrial intelligence.

Life as We Know It (LAWKI)

Astrobiology is based on our current understanding of life on Earth, or as it is sometimes termed, LAWKI—life as we know it. It is clear that our knowledge is therefore limited, and there is no reason why life on other worlds must be identical or even similar to life on our world. On Earth life is based on four basic elements, carbon, hydrogen, oxygen and nitrogen, which are relatively common in the universe today. Carbon is a remarkable atom, which can form unique long chains and rings which serve as the backbone for organic molecules—the basic building blocks of life. In science fiction stories, silicon is sometimes used as the basic component for alien lifeforms, since it lies just below carbon on the periodic table and therefore has some similar properties. Unfortunately, silicon is not believed to be able to easily make complex molecules similar to those that are vital to carbon-based life.

It is necessary to define what we mean by life in the first place. Perhaps the most simplistic definition is that life has the ability to reproduce at the cellular level and evolve. Life on Earth reproduces via DNA (deoxyribonucleic acid), the famous "double helix" molecule. Although there is no reason that life on other worlds should utilize DNA, it is highly probable that some similar type of molecule should be involved. DNA is composed of four different base amino acids, and these and other amino acids make up proteins, important compounds in the function of living organisms. All four of these base amino acids, plus 70 more (including 55 that do not occur naturally in terrestrial life) have been identified in the Murchison meteorite, which landed in Australia in 1972. Amino acids have also been found in the Murray meteorite, as well as in the interstellar medium, leading to the conclusion that the basic building blocks of life are prevalent in the universe today. But amino acids are only the building blocks of life, and the steps between the formation of these chemicals and creation of the first reproducing lifeforms are not definitively understood. This lesson was clearly demonstrated in front of a worldwide audience in the case of the supposed Martian microfossils.

In August, 1996, NASA held a special press conference to announce that signs of ancient life had been found in a meteorite that had been blasted off of Mars. The 3.9-billion-year-old rock, named ALH84001, had been collected in Antarctica in 1984, and like over a dozen other meteorites, had been verified as Martian in origin by the composition of gases trapped within the rock (which matched observations taken from the surface of Mars by the Viking landers). Inside the greenish rock scientists had found what they believed to be four different signatures of life. The first was rosette-shaped nodules of carbonate about 0.05 mm across that were suggested to have been produced by bacteria. The second evidence was tiny grains of magnetite about 50 nm long (less than 1/100th the width of a human hair). Magnetite is a magnetic mineral used by some bacteria on Earth as internal compasses and also produced as a waste product to rid the bacteria of excess iron. The third evidence presented was complex organic molecules called polycyclic aromatic hydrocarbons, or PAHs, which could be a sign of life. The most controversial evidence was tiny tube-shaped objects that some scientists interpreted as microfossils of Martian bacteria.

As soon as the results were announced, scientists from around the world began the process of peer review, rechecking the analysis of the NASA team and searching for alternative explanations and interpretations for the four pieces of evidence. Within several years, the claim of life on Mars had begun to unravel. Logical nonbiological explanations for the carbonate rosettes and magnetite grains were soon published. Although the PAHs were found to be likely Martian in origin and not contamination from the meteorite's time on Earth, the presence of organic molecules does not prove the existence of life (as in the case of other meteorites found on Earth). The supposed microfossils were carefully scrutinized by biologists, with the result that they have been largely refuted as evidence of life (at least as we understand it). The objects were only a few dozen nanometers across, and research strongly suggests that a size at least 10 times that would be necessary to successfully house the "most basic molecular machinery of life" (Kerr 1998, 1398). In comparison, the common E.coli bacteria has a diameter of about 2,000 nm. Therefore, at the time these words are being typed, there is no evidence for past or present life on Mars, although either possibility certainly cannot be ruled out by our limited exploration of the Red Planet.

Since life has not been found on the planet next door, where should we begin our search for the proverbial needle in a haystack? Although humans can only exist within a relatively narrow range of temperatures, and any aquarium owner knows that fish can only thrive within a window of pH levels, there are organisms called extremophiles that can exist under conditions once deemed impossible. For example, thiococcus, a type of bacteria, can actually perform photosynthesis using infrared rather than visible light. For the sake of conversation, we will restrict our search to planets similar to Earth in terms of temperature, composition, and the amount of harmful radiation received from their stars. We will also assume that since it took approximately a billion years for the first life to arise on Earth, and another 3.5 billion years for intelligent life to evolve, the timescale will be similar on other worlds. Therefore we will only look at stars whose lifespan is long enough to allow intelligent life to evolve. Having done this, we are left with spectral class F5 through K5 stars, limiting our search to about 25 percent of the stars in our galaxy. Now that we have narrowed our search to a mere 25 billion stars, is there any way to estimate how many intelligent civilizations await contact?

In 1961, radio astronomer Frank Drake came up with a mathematical expression for this estimate, now called the *Drake equation*. Although it is cited in slightly different forms in different references the concepts are rather basic. There are a number of parameters which must be considered in estimating the total number of intelligent civilizations in our galaxy, and since most of these parameters are not known with certainty, the number of intelligent civilizations can vary from as low as one to as high as several million or more. The various parameters are the number of stars in the Milky Way, the fraction of stars similar to the sun, the fraction of sun-like stars with planets and the number of habitable planets per star, the fraction of such planets where life actually arises, the fraction of those that are inhabited by intelligent life that communicates via modern technology, and the average lifespan of such intelligent civilizations. It is actually this last parameter which has the most variation, since we only have one civilization to use as a model—our own.

Talking to Aliens

Given the uncertainty in the number of candidate civilizations in the Milky Way, and no way to know in which direction to look, astronomers are left with a quandary—do we sit and listen, hoping that our galactic neighbors are talkative, or do we broadcast, like the Whos in Dr. Seuss' classic tale *Horton Hears a Who*, our faint voices shouting to the cosmos "We're here! We're here!" Both tactics have been taken by astronomers over the past few decades. In 1974, Frank Drake and colleagues beamed a message toward the globular cluster M13 in Hercules. It was a digital message which contained basic pictures of a human being, the solar system, and the double helix of DNA.

Most scientists involved in SETI are using a "sit and wait" tactic, hoping to pick up evidence of an intelligent signal. But even this is a monumental task, given the number of possible wavelengths that could be used. Since water is a vital component of life on Earth, one can assume it might also be important for extraterrestrial life forms. Water can be broken into atomic oxygen (O) and the hydroxide radical (OH), which are called the dissociation products of water. Each has distinctive spectral lines, the most important of which lie between frequencies of 1400 and 1700 MHz. In 1976 physicist Bernard Oliver suggested that searches for extraterrestrial intelligence should focus on this range of frequencies, which was dubbed the "water hole." It was argued that intelligent species would meet around this technological water hole just as creatures habitually do on Earth around more literal water holes. Carl Sagan suggested that since the 21-cm radiation of cool hydrogen should be known to any technological society, multiples of that frequency (1420 MHz) might be a good choice for a broadcast, especially frequencies such as $1420 \times \pi$ or $1420/\pi$. This is the assumption he used in his novel *Contact.*

Besides radio signals, humans have also sent physical artifacts into space that can tell extraterrestrial travelers about our planet. For example, the Pioneer 10 and 11 spacecraft that traveled by Jupiter in the 1970s were slingshotted out of the solar system by the giant planet's gravity. Although they are not slated to pass by any known stars in the foreseeable future, each has a gold-plated plaque designed by Carl Sagan and Frank Drake which tells the simple story of the species that sent them. Next to a diagram of the spacecraft (shown for scale) are the naked forms of a male and female human, the male's hand raised in a gesture of friendship. A diagram of the solar system with Earth highlighted is also included, along with several other pictograms. Several years later the two Voyager spacecraft were launched on a tour of the giant planets and beyond. Each contained a gold-plated record and materials (and instructions) to build a record player. As Carl Sagan explained, each contained

Greetings in sixty human tongues, as well as the hellos of the humpback whales. We sent photographs of humans from all over the world caring for one another, learning, fabricating tools and art and responding to challenges. There is an hour and a half of exquisite music from many cultures, some of it expressing our sense of cosmic loneliness, our wish to end our isolation, our longing to make contact with other beings in the Cosmos. (1980, 287)

Is this search in vain? Is there anyone out there? If so, could they really understand the messages we have sent? Would they care enough to answer us? Only one thing is certain—nothing is truly in vain in science if it allows us to improve our technology and expand our knowledge of the universe, even if that knowledge leads us to the conclusion that we are alone. In the words of Carl Sagan, "Many, perhaps most, of our messages will be indecipherable. But we have sent them because it is important to try" (Ibid.).

CONCLUSION

We have now reached the end of our journey through space and time, from the origin of the universe to its demise, from our solar system to the faintest and most distant galaxies and quasars. It is certainly impressive how much we know about the universe we live in, given our fixed position in space and time within it. We began with the concept that the earth was the center of the cosmos, and have been left with the idea that we are not only denied a central position, but we are not even made of the most prevalent materials in the universe (namely dark matter and dark energy). If eternal inflation is correct, our visible universe is also one of perhaps an infinite number in a larger multiverse, further removing us from a place of importance. This concept was called "the extreme Copernican Principle" by cosmologist David Schramm.

Despite all that we know about the universe, there remain serious unanswered questions, including the nature of dark energy, the properties of the correct grand unified theory and quantum gravity model, and the identity of the inflaton field. Once gravity waves are directly detected, inflationary theories may be further bolstered and the ekpyrotic model refuted, or perhaps vice versa. The scientific method is a cruel mistress, and no matter the beauty of the theory, it can fall victim to the blades of observation and experimentation in the flash of an eye.

Never was the scientific method more evident to the general public than in the case of Pluto. When I began writing this book, Pluto was classified as a planet. By the time I had finished, the International Astronomical Union had reclassified it as a dwarf planet, and the news was awash with debate over the decision, much of it emotionally charged. The response of many people to the news of Pluto's "demotion" from planetary status points out a drastic misunderstanding in today's culture of the nature of science. As we have seen, science is the investigation of the natural world, a never-ending search for answers to current questions about how the universe works, as well as the search for new and better questions to ask. Sometimes discoveries are made and not completely understood at the time, leading to erroneous assumptions being made. Unfortunately, as we have seen, even in science misconceptions can be stubborn and are not always swift to be remedied. Such is the case with Pluto.

Decades after its discovery, astronomers had learned more about Pluto's properties, and it was obvious that it was not only very different from the four Earth-like rock and metal planets and the four Jupiter-like gas giant planets, but that it shared more in common with the myriad icy bodies in the outer solar system. The suggestion was originally made by those in the International Astronomical Union (IAU) that Pluto be cross-classified as a planet and Kuiper Belt object, a member of the icy asteroid belt of sorts beyond the orbit of Neptune. The outcry from the general public was surprising, and rather alarming. As you have learned, scientists admit their mistakes all the time, reclassifying objects, acknowledging that experiments and observations were erroneous, even acknowledging the occasional embarrassing error in a theoretical calculation. All of this passes without a blink from John Q. Public. Where was the outcry several years ago when the IAU extended the classification of stars to include the new classes L and T? Astronomy books had to be changed and students scrambled to create a new mnemonic device for the extended classification system (if anyone has a really good one for OBAFGKMLT, I'm all ears). When rabbits were reclassified from rodents to lagomorphs in the early 1900s, the Easter bunny did not suddenly change his trademark confectionary gifts.

Why is Pluto so different? Pluto's proposed reclassification brought claims that textbooks would need to be changed, and that children would have to learn new facts. Yes, they will, and they have, throughout the history of science! Astronomical discoveries happen at such a fast and furious rate these days that textbooks are out of date before their ink is even dry. The number of crossed-out sentences and hastily scribbled additions to the margins of my lecture notes drives home that point. When classes start every semester, I preface my comments on the first day of class with the warning that 10 percent of what I will teach the students may eventually be proven wrong (a rather conservative estimate). As members of a technological society, we have had to learn to deal with constant change, which is admittedly not always comfortable. I still have a box of eight-track tapes in my basement, for reasons that have nothing to do with science. Perhaps Pluto's demotion hits some people harder than others for a similar reason. But as Max Tegmark hopefully offers, "Perhaps we will gradually get more used to the weird ways of our cosmos, and even find its strangeness to be part of its charm" (2004, 490).

It is this very same ever-changing nature of scientific knowledge which assures that there will always be questions to ask and experiments to run. There is always that one further test which might prove a long-cherished theory wrong, or point the way to an improvement in our current understanding (as in the case of inflation and the big bang theory). Science does not belong to the scientists, and the universe does not belong to cosmologists. We are all citizens of this grand and infinite cosmos, our bodies made of the very atoms created in the nuclear furnaces of long-dead stars, our minds free to soar through space and time to the very beginning of the universe. Take the time to go out on the next clear night, away from the distracting lights of modern society, and take a moment to gaze upward at our stellar cousins, to whom we owe our very existence. In the words of Annie Jump Cannon, who became intimately familiar with more than 350,000 stars during her distinguished career, "starlovers and stargazers are linked together in bonds of friendship while the same sky overarches us all, and we do our best to increase the sum of human knowledge as pertains to the Story of Starlight" (1941, 61).

GLOSSARY

Absolute magnitude.   How bright a star would appear at a distance of 10 pc.

Accretion disk.   A disk of hot material spiraling onto a stellar corpse, such as a white dwarf.

Active galactic nuclei.   A unified explanation for QSOs, Seyfert galaxies, radio galaxies, and blazars based on the interactions between a supermassive black hole and its local environment.

AdS/CFT.   The concept that a quantum theory in a higher-dimensional space can be modeled by another theory that inhabits the boundary of that space.

Angular power spectrum.   A graph of the cosmic microwave background showing the relative strength of temperature variations of various angular sizes.

Apparent magnitude.   The observed brightness of a star as seen from Earth.

Astronomical unit.   The average sun-Earth distance (150 million km).

Asymptotic branch.   The low temperature, high luminosity section of the HR diagram that dying stars of solar-type mass inhabit after ceasing the triple-alpha cycle.

Barred spiral.   A spiral galaxy whose center features a bar-like structure.

Baryogenesis.   The physical processes in the early universe that selected out matter over antimatter.

Baryonic matter. Normal matter composed of protons and neutrons.

Big bang. The scientific theory that the universe began in a hot, dense state and has expanded ever since.

BL Lac object. A class of powerful and highly variable nuclei of galaxies.

Black hole. An object whose gravitational field is so intense that not even light can escape.

Blackbody. A hypothetical body that is a perfect absorber and perfect emitter.

Blazar. A term for BL Lac objects and optically violent variable quasars.

Boson. A subatomic particle that does not obey the Pauli exclusion principle, such as the photon.

Brane. An object with two or more spatial dimensions predicted by M-theory.

Brown dwarf. An object more massive than a planet but whose mass was too low to become a star.

Carbon cycle. The chain of hydrogen fusion reactions used by massive main sequence stars to generate energy, where carbon acts as a catalyst.

Cepheids. Unstable stars whose period of variation depends on their average luminosity.

Closed universe. A universe that will collapse in the future because the density of matter and energy is above the critical density.

Cold dark matter. Dark matter that moves much slower than the speed of light and clumps together to seed galaxy formation.

Cosmic distance ladder. A series of corroborating methods used to estimate distances to stars and galaxies.

Cosmic magnification. A weak variety of gravitational lensing that brightens the images of distant galaxies.

Cosmic microwave background. A pervasive leftover energy from the early universe which has significantly cooled over time to the current 2.7 K.

Cosmic shear.   A weak variety of gravitational lensing that distorts the shapes of distant galaxies.

Cosmic string.   A one-dimensional defect in space-time predicted by some GUTS.

Cosmological constant.   A constant term in Einstein's field equations of general relativity, originally conceived by Einstein to balance gravity and keep the universe static, now used to describe a leading candidate for dark energy.

Cosmological principle.   On average, the universe looks the same viewed from any location and in any direction.

Cosmology.   The scientific study of the composition, structure, and evolution of the universe.

Critical density.   The density of matter and energy in the universe required for the expansion of the universe to be exactly balanced by gravity.

Dark age.   The period of time between recombination and the first generation of stars when the universe was dark.

Dark energy.   A generic name for the unknown cause of the current accelerating expansion of the universe.

Dark matter.   Invisible material whose presence is discovered through its gravitational effect.

Deceleration parameter.   A measure of the rate of change of the expansion of the universe.

Deferent.   In the Ptolemaic cosmology, the main circle comprising a planet's orbit around Earth.

Degeneracy pressure.   The pressure created by the Pauli exclusion principle that prevents further collapse of a stellar corpse.

Density perturbation.   Slight variations in the density of matter that were created in the early universe and led to the formation of structures such as galaxies.

Density wave theory.   Spiral arms form when gas, dust, and stars slow their revolution around the galaxy and temporarily clump together like a traffic jam.

Distance modulus formula.   The distance to an object is determined by comparing its apparent and true brightness.

Doppler broadening.   The widening of a spectral line due to the rotation of an object along our line of sight.

Doppler effect.   Wavelengths of sound or light lengthen as the source moves away from the observer and shorten as the source approaches.

Drake equation.   A mathematical estimate of the number of technological societies in the universe

Duality.   A connection between two physical theories that demonstrates that they give the same results.

Einstein field equations.   The fundamental equations of general relativity that describe how matter and energy warp space-time.

Ekpyrotic model.   An alternative to inflation in which the universe is created by the collision of two branes.

Electromagnetic spectrum.   The continuum of all possible wavelengths of light (electromagnetic waves).

Electroweak force.   The high-energy unification of the electromagnetic and weak nuclear forces.

Elliptical galaxy.   A galaxy with a round to elliptical shape and no spiral arms.

Epicycle.   In the Ptolemaic cosmology, the small circle on which a planet is attached, which in turn is attached to the deferent.

Equant.   In the Ptolemaic cosmology, the center of uniform motion of a planet's orbit.

Eternal inflation.   The concept that once inflation begins it will never cease.

Extinction.   Interstellar dust makes distant objects appear artificially dim.

False vacuum.   A local minimum value of energy that is not the overall lowest possible energy state.

Fermion.   A subatomic particle that obeys the Pauli exclusion principle, such as the electron.

Field.   A physical quantity whose properties can be defined at a particular location.

Final anthropic principle.   The suggestion that once life arises in the universe it can never be extinguished.

Flat universe.   A universe that has achieved the critical density.

Friedmann-Robertson-Walker model.   A mathematical model of the universe that can be open, closed, or flat (in a geometrical sense).

Geocentric universe.   An ancient model of the universe with Earth at the center.

Globular cluster.   A dense, spherical mass of hundreds of thousands of older stars.

Grand unified theories (GUT).   A class of models that seek to unite the strong nuclear and electroweak forces.

Gravitational lensing.   Multiple images of a distant object (such as a QSO) caused by the gravitational field of an intervening object (such as a galaxy).

Gravity waves.   Ripples in space-time generated by the motion of massive objects.

Hadron.   A particle that interacts via the strong nuclear force.

Half-life.   The time required for half of a radioactive sample to decay.

Heisenberg uncertainty principle.   In quantum mechanical systems, it is impossible to know both the location and motion of a particle, or the duration and energy of an event, simultaneously to absolute certainty.

Heliocentric universe.   A model of the universe that places the sun at the center.

Hertzsprung-Russell (HR) diagram. A plot of the spectral class (temperature) and luminosity (absolute magnitude) of stars.

HI region. A cool hydrogen cloud with a distinctive radio emission at 21 cm.

Hierarchy problem. The puzzling fact that gravity is so much weaker than the other forces.

Higgs field. A quantum mechanical field supposed to make subatomic particles massive rather than massless.

HII region. A hot cloud of glowing hydrogen.

Horizontal branch. The section of the HR diagram between the red giant branch and main sequence where stars undergoing core helium burning are located.

Hot dark matter. Dark matter that travels near the speed of light and can only form very large structures.

Hubble tuning fork diagram. A pictorial representation of the various classes and subclasses of galaxies.

Hubble's law. The more distant a galaxy, the faster it appears to recede, evidence for the expansion of the universe.

Hypothesis. A proposed scientific explanation for some aspect of the physical world.

Inflation. A class of models in which our observable universe underwent an early period of exponential expansion.

Inflationary ΛCDM model. Also called the concordance model, a combined model of cold dark matter, dark energy, and inflation that is supported by many observations.

Interstellar reddening. Interstellar dust scatters the blue component of starlight, making the star appear less blue (or "redder").

Island universe hypothesis. The suggestion in the late 1700s that there might be other galaxies outside the Milky Way.

Keplerian rotation curve.   The rapidly declining rotation curve generated by a system where the mass is concentrated in the center (like the solar system).

Lenticular galaxy.   A lens-shaped galaxy resembling a spiral without arms.

Local Group.   The several dozen galaxies clustered around the Milky Way and Andromeda Galaxy.

Magnetic monopole.   A hypothetical particle that represents an isolated north or south magnetic pole.

Main sequence.   The diagonal line on the HR diagram that represents normal stars creating energy by fusion of hydrogen in their core.

Main sequence turnoff.   The position of the top of the main sequence on an HR diagram for a star cluster, determined by the age of the cluster.

Malmquist bias.   Dim members of any population will be overlooked in surveys of distant objects.

Metallicity.   The percentage of a star (by mass) composed of elements heavier than helium.

Microlensing.   A point-like object passes in front of another point-like object (such as stars), focusing the light and temporarily brightening the system.

M-theory.   A speculative theory invoking extra dimensions that includes superstrings and supergravity as special cases.

Multiverse.   A collection of pocket universes (or other multiple universes).

Nebula.   A cloud of gas and/or dust in space.

Neutron star.   A high density, medium-mass stellar corpse made of neutrons.

Non-Hubble flow.   Also called peculiar motion, motion of galaxies independent of the expansion of the universe.

Nova.   Originally a generic term for any exploded star, more correctly the outburst of a white dwarf in a binary star system.

Nuclear cosmochronometry.   The use of the decay of radioactive elements in old stars to estimate their age.

OB association.   A loose aggregation of very hot, young stars.

Olbers' paradox.   The principle that if the universe were eternal and infinite the night sky could not be dark.

Open cluster.   An irregular group of dozens to hundreds of young stars.

Open universe.   A universe whose density is below the critical density that will expand forever.

Pair-instability supernova.   The implosive death of the supermassive first generation of stars.

Parallax.   The apparent shift in the positions of nearby stars relative to more distant stars due to the motion of Earth around the sun.

Parsec.   The distance of a star whose parallax would be one arc second (3.26 light years).

Participatory anthropic principle.   The idea that the events in the universe are not actualized until observed, necessitating the existence of intelligent life.

Pauli exclusion principle.   Identical fermions cannot occupy the same quantum state at the same time.

Perfect cosmological principle.   The universe has always looked the same, everywhere and at all times.

Period-luminosity relationship.   The relationship between average luminosity and period of pulsation for Cepheid variables, used to determine distances to galaxies.

Planck curve.   A graph showing the intensities of a range of wavelengths of light for an object of a particular temperature.

Planck era.   The first 10^{-43} seconds of the universe, when all four forces were unified.

Planetary nebula.   The outer layers of a dying star puffed off into space.

Pocket universe.   Separate universes independently created from the false vacuum.

Polarization.   A property of light that describes whether the directions of oscillation of different waves are aligned or not.

Population I.   Stars like the sun that are composed of recycled interstellar material with higher metallicities.

Population II.   Older stars that represent a previous generation and are relatively lower in elements heavier than helium.

Population III.   The still-unobserved first generation of stars predicted to be made of pure hydrogen and helium.

Primordial nucleosynthesis.   The processes that generated helium in the first few minutes of the universe.

Proton–proton cycle.   The chain of hydrogen fusion reactions used by solar-type main sequence stars to generate energy.

Protostar.   An object in the process of becoming a star.

Pulsar.   A rapidly rotating neutron star detected by its beams of radio waves.

Quantum chromodynamics.   The quantum theory of strong nuclear force interactions.

Quantum electrodynamics.   The quantum theory of electromagnetic interactions.

Quantum gravity.   Any scientific model that attempts to unite general relativity and quantum mechanics.

Quantum mechanics.   The principles that explain the behavior of microscopic particles such as atoms.

Quantum tunneling.   The ability of a quantum mechanical system to move from one side of an energy barrier to another without having enough energy to pass over the barrier.

Quark.   A fundamental particle that comprises neutrons and protons.

Quasi-stellar object (QSO). A generic term for star-like, highly energetic galactic nuclei of high redshift, including quasars.

Quintessence. A type of dark energy that varies with time.

Radial velocity. Motion of an object toward or away from an observer.

Radio galaxy. A galaxy with unusually strong radio emissions.

Radiometric dating. Utilizing radioactive decay to estimate the age of an object.

Recombination. When the universe cooled to a few thousand degrees K, electrons were finally bound to nuclei to form stable atoms.

Red giant branch. The low temperature, high luminosity section of the HR diagram inhabited by stars undergoing hydrogen shell burning.

Redshift. The wavelength of light emitted by a receding object appears lengthened.

Rotation curve. A graph that plots the orbital speeds of objects located at different distances from the center of a galaxy.

r-process. The "rapid" neutron capture process by which supernovae create nuclei heavier than iron.

RR Lyrae stars. Unstable, low mass stars that pulsate within a day or less.

Scientific method. A continual process of hypothesis, observation, and experimentation used by scientists to understand the world.

Seyfert galaxy. A spiral galaxy with an unusually bright core and other unusual emissions.

Singularity. In general relativity, a situation where the field equations break down, such as at the center of a black hole or the origin of the universe.

Space-time. The fabric of the universe composed of three dimensions of space interwoven with one dimension of time.

Spectral classes. The classification of stars by differences in their spectra, due to different surface temperatures.

Spectroscopic parallax.   An HR diagram of stars of known luminosity and spectral class can be used to estimate the luminosity, and from that distance, for other stars.

Spectroscopy.   The process of separating a beam of light into its constituent wavelengths for analyzing.

Spiral galaxy.   A galaxy that has two or more arms composed of gas, dust, and stars.

Spiral nebula.   The term used for spiral galaxies before the 1940s.

Spontaneous symmetry breaking.   A system automatically moves from a state of perfect symmetry to a state of broken symmetry.

s-process.   The "slow" neutron capture process used by red giants to create nuclei heavier than iron.

Standard candle.   Any astronomical object whose true brightness can be estimated with some certainty.

Standard model.   The current description of particle physics interactions, including QCD and the electroweak theory.

Steady-state theory.   A out-dated model of the universe as an eternal, unchanging system.

Stefan-Boltzmann law.   The mathematical relationship relating the luminosity of a star to its size and temperature.

Stellar nucleosynthesis.   The processes by which stars create atomic nuclei through fusion.

String theory.   The concept that what are observed as elementary particles are different modes of vibration of tiny string-like objects.

Strong anthropic principle.   The suggestion that the universe must be fine-tuned to allow for the existence of life.

Strong nuclear force.   The strongest of the four fundamental forces, which describes how quarks interact to form protons and neutrons.

Supercooling.   A system cools to a critical temperature and does not automatically undergo the appropriate transition, such as in freezing rain.

Supergravity.   A theory that unites supersymmetry and general relativity.

Supermassive black hole.   A black hole millions of times more massive than the sun, found at the heart of a galaxy.

Superstring.   A string theory that unifies bosons and fermions.

Supersymmetry.   A generic term for models that unifies bosons and fermions.

Synchrotron radiation.   A characteristic radio signal created by electrons moving in a magnetic field.

Theory.   A hypothesis that has been rigorously tested and is generally considered among the best current explanations.

Triple-alpha cycle.   The fusion cycle by which carbon and oxygen are created from helium nuclei.

Type Ia supernova.   The explosion of a white dwarf.

Type II supernova.   The explosion of a red supergiant.

Weak anthropic principle.   The concept that if the universe were different, we would not be here to study it.

Weak nuclear force.   The fundamental force that mediates radioactive decay.

Weakly interacting massive particle (WIMP).   Theorized particles of cold dark matter, including supersymmetric partners and axions.

White dwarf.   A compact stellar corpse about the mass of the sun and the diameter of Earth.

Wien's law.   The hotter the star, the shorter the wavelength at which it emits most of its light, making it appear bluer.

Zone of avoidance.   Dense dust and gas in the plane of the Milky Way impede observations in that portion of the sky.

SELECTED BIBLIOGRAPHY

Adams, Fred C. and Gregory Laughlin. 1999. *The Five Ages of the Universe*. New York: Free Press. This is the seminal popular-level work on the fate of the universe, written by the researchers themselves. A thought-provoking (and sometimes depressing) survey of the ultimate fate of matter itself, long after the demise of the solar system.

Alpher, Ralph A. and Robert Herman. 2001. *Genesis of the Big Bang*. Oxford: Oxford University Press. Two of the main characters in the development of the big bang theory recount their roles and the work of famed physicist George Gamow. The personal nature of this text is a unique window into the world of science and scientists.

arXiv preprint archive (http://arXiv.org). Given that it can take a year for scientific papers to be published in professional journals, scientists have taken to posting "preprints" of their papers in online archives. arXiv is the largest and best known such archive in the physics and astronomy communities. The interested reader can search for papers by subject, and will find the most up-to-date research on a variety of cutting-edge topics.

Asimov, Isaac. 1956. The Last Question. Available at http://www.multivax.com/last_question.html (accessed September 1, 2006). This classic short story may seem a little dated, but it is an excellent overview of the concept of the fate of the universe, written by a master of science fiction.

Baade, Walter. 1956. The Period-luminosity Relation of the Cepheids. *Publications of the Astronomical Society of the Pacific* 68: 5–16. A transcript of a lecture given in honor of his receiving the Bruce Medal from the ASP, this article is an excellent and readable summary of the history of the calibration of the Cepheid period-luminosity relationship.

Barrow, John D. and Frank J. Tipler. 1988. *The Anthropic Principle*. Oxford: Oxford University Press. This technical book is considered the seminal work on the anthropic principle. The interested and dedicated reader will find this work challenging but fascinating.

Bromm, Volker. 2003. Cosmic Renaissance. *Mercury* 32(5): 25–33. Bromm, one of the active researchers in the field of population III stars, presents a highly readable survey of current theories of the properties of the first generation of stars.

Burbidge, E. Margaret, G. R. Burbidge, William A. Fowler, and F. Hoyle. 1957. Synthesis of the Elements in Stars. *Reviews of Modern Physics* 29(4): 547–650. The so-called B^2FH paper demonstrated that all elements heavier than hydrogen could be created through the lives and death of stars, results that have largely remained the same to this day. One of the most important papers in twentieth-century physics, this lengthy and detailed work is largely meant for a technical audience.

Cannon, Annie J. 1941. The Story of Starlight. *The Telescope* 8(3): 56–61. This article is the transcript of a popular-level radio talk by the woman who single-handedly classified more stellar spectra than any other human being. Her overview of the importance of spectroscopy is written in a comfortable, personal style, and her genuine affection for her friends, the stars, shines throughout.

Crowe, Michael J. 1994. *Modern Theories of the Universe from Herschel to Hubble.* New York: Dover. Crowe surveys the development of modern cosmology through the multiple lenses of science, history, and philosophy. This text is a valuable balance of excerpts from original texts and insightful commentary.

Davies, P. C. W. and J. Brown, eds. 1988. *Superstrings: A Theory of Everything?* Cambridge: Cambridge University Press. Although somewhat dated, this is a valuable review of the early history of string theory. Much of the work is a series of interviews with the important players in string theory in the 1980s, who provide their own personal insight into the controversial theory.

Einstein, Albert. 1961. *Relativity: The Special and General Theory*, 2nd ed. New York: Crown. Although intended as an overview of both the special and general theories of relativity for the general reader, it is not an easy read. It is valuable as an explanation of Einstein's seminal work through his own words, including unique insights into his thinking.

Flamsteed, Sam. March 1995. Crisis in the Cosmos. *Discover*, 66–77. Despite its sensational title, this popular-level article covers Wendy Freedman's results concerning the Hubble constant and other unsettling observations clearly and with brutal honesty. It paints a vivid example of the scientific method in action in cosmology.

Freedman, Wendy L., Barry L. Madore, Jeremy R. Mould, et al. 1994. Distance to the Virgo Cluster Galaxy M100 from Hubble Space Telescope Observations of Cepheids. *Nature* 371: 757–762. This classic paper affirmed that the value of the Hubble constant could not be accommodated with the accepted ages of globular clusters in a flat universe without a cosmological constant. The immediate result of the publication of this paper was a flurry of popular-level, sensationalized articles declaring the big bang model to be in trouble, causing some scientists to consider the possibility that the cosmological constant might not be zero after all (anticipating the discovery of dark energy). Although written for a technical audience, the writing style is clear and carefully lays out how their experiment was done and their conclusions reached, making this an excellent example of the scientific method in action.

Freedman Wendy L., Barry F. Madore, Brad K. Gibson, et al. 2001. Final Results from the Hubble Space Telescope Key Project to Measure the Hubble Constant. *Astrophysical Journal* 553: 47–72. As the title explains, this is the final report of Freedman's team concerning the value of the Hubble constant. Although this detailed paper was intended for their professional colleagues, the interested reader will get a sense of the excruciating care taken by astronomers in collecting their data.

Gamow, George.The Evolutionary Universe. September 1956. *Scientific American*, 136–165. A nontechnical review of the cosmological models leading up to the development of the big bang, written by one of the theory's founders. Gamow finishes with a summary of his own model, which he interestingly does not call the big bang (due to the term's original pejorative nature). An important popular-level paper in the history of modern cosmology.

Gefter, Amanda. 2003. Decoding the Mystery of Dark Energy. *Mercury* 32(5): 34–40. Various models of dark energy and dark matter are reviewed at a popular-level. This is a highly recommended first introduction to dark energy for readers of limited science background.

Geller, Margaret J. 1990. Mapping the Universe: Slices and Bubbles. *Mercury* 19(3): 66–76. Geller leads the nontechnical reader through a brief history of mapping our own world before discussing her seminal research in the mapping of structure in the universe. Her personal account of the groundbreaking discovery of the bubble-like system of filaments and voids makes this an important reference.

Goldsmith, Donald and Tobias Owen. 2002. *The Search for Life in the Universe*, 3rd ed. Sausalito, CA: University Science Books. Touted as the "premiere text for courses dealing with astrobiology," this work lives up to its press. Basics of astrophysics and biology are surveyed, as well as aspects of interstellar communication and travel. This text is highly recommended for all readers interested in life beyond Earth and the problems of interstellar communication.

Gould, Stephen J. 1999. *Rocks of Ages*. New York: Ballantine. Respected paleontologist and popular-level science writer Stephen Jay Gould investigates the relationship between science and religion and finds a place of peace in his concept of nonoverlapping magisteria (NOMA). This work is strongly recommended for anyone interested in the interface between theology and cosmology, or science and religion in general.

Guth, Alan. 1997. *The Inflationary Universe*. Reading, MA: Addison-Wesley. This engaging first-person account of the development of the inflationary model is a must-read for any cosmology enthusiast. Guth's humor and humility provide a refreshing perspective on the process of science and the sometimes thorny relationships between scientists.

Gyatso, Tenzin (Dalai Lama). 2005. *The Universe in a Single Atom*. New York: Morgan Road Books. The widely recognized spiritual leader of Tibetan Buddhism and Nobel Peace Prize laureate demonstrates that science and religion cannot only coexist, but can enrich the experience of the other. Although written for a popular audience, some of the Buddhist philosophy may be confusing to the casual reader.

Harrison, Edward. 1987. *Darkness at Night*. Cambridge, MA: Harvard University Press. Harrison delivers a detailed but accessible survey of the development, history, and solution of Olbers' paradox. Although narrow in focus, it is the seminal survey of this topic.

———. 2000. *Cosmology*, 2nd ed. Cambridge: Cambridge University Press. Known for his deep philosophical musings about cosmology, Harrison weaves together historical developments in science with an insightful reflection as to their meaning.

Although it sometimes reads like a textbook, it is such a unique work that it belongs on the bookshelf of every serious devotee of cosmology.

Hawking, Stephen W. 1993. *Black Holes and Baby Universes and Other Essays.* New York: Bantam Books. This popular-level collection of essays is a much easier read than Hawking's famous *A Brief History of Time*, and provides an overview of both Hawking's research into black holes and Hawking the man, including his physical struggles with ALS.

————, ed. 2002. *On the Shoulders of Giants.* Philadelphia: Running Press. Hawking edited this collection of the most seminal works of Copernicus, Kepler, Galileo, Newton, and Einstein and wrote brief but informative overviews of each scientist's life. This thick tome provides the most important works of the Copernican revolution in one volume, as well as a selection from the work of Einstein, who overthrew the mechanistic Newtonian universe after nearly 400 years of unrivalled success. This work is recommended for the reader interested in exploring the Copernican Revolution through primary texts.

Hawking, Stephen and Roger Penrose. 1996. *The Nature of Space and Time.* Princeton: Princeton University Press. This volume is based on a series of alternating lectures given by the two famed theoretical physicists on the nature of the interplay between cosmology and quantum mechanics. Humor and occasional mathematics are intertwined, making this a unique work. It is recommended for the serious reader.

Heath, Thomas L. 1991. *Greek Astronomy.* New York: Dover. First published in 1932, this work begins with a detailed introduction to the contributions of early Greek astronomers. The remainder of the text is judiciously selected excerpts from primary texts, some not easily accessible elsewhere.

Hetherington, Norriss, ed. 1993. *Cosmology.* New York: Garland Publishing. The secondary title of this work "Historical, Literary, Philosophical, Religious, and Scientific Perspectives" accurately summarizes the importance of this collection of essays. Unlike similar volumes, this work includes cosmologies from non-Western traditions. Highly recommended for the reader interested in the influence cosmology has had on other aspects of human society.

Hubble, Edwin. 1929. A Relation Between Distance and Radial Velocity Among Extra-galactic Nebulae. *Proceedings of the National Academy of Sciences* 15: 168–174. Although not the first scientific paper to announce the linear relationship between distance and redshift, Hubble's paper was the seminal work on the topic, and earned the relationship his name. It was clearly written for a technical audience, so its value is largely historical.

Inglis, Mike. 2003. *Observer's Guide to Stellar Evolution.* London: Springer. Inglis offers an overview of spectral classification and stellar evolution for amateur astronomers. Although meant for the serious observer of the night sky, the nontechnical reader will find this a useful survey of important concepts.

Jones, Bessie Zaban and Lyle Gifford Boyd. 1971. *The Harvard College Observatory.* Cambridge, MA: Harvard University Press. This detailed history of the HCO through 1919 includes a frank discussion of the role of women in the early twentieth century (including Antonia Maury, Annie Jump Cannon, Williamina Fleming, and Henrietta Leavitt). Highly recommended for anyone interested in the important contributions of these women and the conditions under which they worked.

Jones, Mark H. and Robert J. A. Lambourne, eds. 2004. *An Introduction to Galaxies and Cosmology.* Cambridge: Cambridge University Press. A textbook designed for lower-level undergraduates, this work provides an excellent background for the reader interested in galactic astronomy. The visual learner will appreciate the plentiful diagrams and illustrations.

Kane, Gordon. 2002. Anthropic Principles. *Phi Kappa Phi Forum* 82(4): 21–24. Kane addresses the apparent fine-tuning of the universe in accessible language. This article is a useful summary of various scientific explanations, including the anthropic principle and multiverse theory.

Kerr, Richard A. 1998. Requiem for Life on Mars? Support for Microbes Fades. *Science* 282: 1398–1400. This concise article demonstrates the scientific method in action, as evidence for fossil life on Mars was critiqued by the scientific community and found wanting. This is a valuable summary of an event that created quite a stir in both scientific and media circles a decade ago, but has been largely forgotten since.

Kirshner, Robert P. 2002. *The Extravagant Universe.* Princeton: Princeton University Press. Harvard's Kirshner played a lead role in the discovery of the accelerating universe. An excellent public speaker, Kirshner's style is engaging and reflects the excitement of the process of discovery in science.

Kolb, Edward W. and Michael S. Turner. 1990. *The Early Universe.* Redwood City, CA: Addison-Wesley. Two leading particle physicists review the interface between their field and cosmology (circa 1990) for a technical audience. Although dense with mathematics, the text is clearly written and punctuated by the authors' wit. A companion volume contains reprints of seminal papers in particle physics and cosmology.

Koopmans, Leon V. E. and Roger D. Blanchard. 2004. Gravitational Lenses. *Physics Today* 57(6): 45–51. This article is an overview of various aspects of gravitational lensing including microlensing, written at an introductory level. Theory is kept to a minimum, as the article's main thrust is the important applications of lensing and the breadth of information that can be obtained.

Kragh, Helge. 1996. *Cosmology and Controversy.* Princeton: Princeton University Press. Kragh's seminal work is a detailed study of the development of the big bang theory and its eventual victory over the rival steady-state. Written by a historian of science, it presents a different point of view compared to similar works written by scientists. Densely referenced, it gives the interested reader an impressive variety of avenues for further exploration.

Krauss, Lawrence. 1999. Cosmological Antigravity. *Scientific American* 280(1): 34–41. Krauss presents an overview of the history of the cosmological constant, and the reasons for its resurgence (the "age crisis" and the accelerating universe). Written by an important researcher in the fields of globular cluster ages and particle physics, this paper is important not only for its accessibility but its timeliness (published at the time of the first evidence for the accelerating universe).

———. 2005. School Boards Want to 'Teach the Controversy.' What Controversy? *Physics and Society* 34(3): 9. Written in response to various attempts to include "Intelligent Design" in the science classroom, physicist Larry Krauss clearly lays out his own support for a reasonable coexistence between science and religion, utilizing specific examples from both. This insightful editorial is highly recommended.

Leavitt, Henrietta S. 1912. Periods of Twenty-five Variable Stars in the Small Magellanic Cloud. *Harvard College Observatory Circular* 173: 1–3. This brief work is simply one of the most important astronomical papers of the early twentieth century. Leavitt clearly lays out her argument and observations, laying the groundwork for the most successful method of distance determination in cosmology. The historical significance of this paper (including the fact that a woman was the sole author) cannot be overly stressed.

Lemonick, Michael D. Before the Big Bang. February 2004. *Discover*, 35–41. This article contains a popular-level discussion of the ekpyrotic and cyclic models. Quotes from both proponents and skeptics set the models within the larger process of the scientific method.

Lemonick, Michael D. and J. Madeleine Nash. Unraveling Universe. March 6, 1995. *Time*, 76–84. One of the many popular-level articles published in the wake of Wendy Freedman's confirmation that the Hubble constant was in conflict with a flat, zero-cosmological constant universe. Both sides of the issue are represented, as well as the difficulties inherent in making such observations. A flavor of the sometimes colorful personalities involved gives this article some spice.

Lightman, Alan and Roberta Brawer. 1990. *Origins*. Cambridge, MA: Cambridge University Press. This text is a valuable collection of interviews with famous cosmologists and particle physicists, including several Nobel laureates. It balances scientific opinions and insights with unique literary snapshots of the physicists as human beings.

Macpherson, Hector. 1919. The Problem of Island Universes. *The Observatory* 42(543): 329–334. Macpherson offers an excellent and timely overview of cosmology in the early twentieth century written for a popular-level audience. It is highly recommended for readers interested in the tension in cosmology during this period of time.

Martínez, Vicent, Virginia Trimble, and María Jesús Pons-Bordería, eds. 2001. *Historical Development of Modern Cosmology*. San Francisco: Astronomical Society of the Pacific. These proceedings from a summer school on the history of modern cosmology contain a rich variety of articles written by some of the heavy-hitters in the field as well as lesser-known researchers. Most articles are easily accessible for the popular-level reader, and provide information not usually included in introductory trade books.

Munitz, Milton K., ed. 1957. *Theories of the Universe*. Glencoe, IL: The Free Press. Munitz has assembled a valuable collection of excerpts from primary texts that reflect cosmology from ancient cultures through the early days of the big bang/steady-state debate, with few editorial comments. This work is a must-have for all those interested in reading about the early history of cosmology through original voices.

Naeye, Robert. Delving into Extra Dimensions. June 2003. *Sky and Telescope*, 38–44. This article is a clearly written, nontechnical survey of theories incorporating extra dimensions, including several different brane worlds. Candid comments by supporters and skeptics provide a balanced approach to these speculative theories.

Peebles, P. J. E. and Bharat Ratra. 2003. The Cosmological Constant and Dark Energy. *Reviews of Modern Physics* 75: 559–606. This work is a scholarly review of the history of the concept of the cosmological constant as well as current observations and theoretical models. Despite its appearance in a scholarly journal, the paper contains relatively little mathematics and is recommended for the interested reader.

Perivolaropoulos, L. 2005. The Rise and Fall of the Cosmic String Theory for Cosmological Perturbations. Available at http://arxiv.org/PS_cache/astro-ph/pdf/0501/0501590.pdf (accessed September 2, 2006). This paper presents a review of the history of cosmic strings and their implications for cosmology. Although written for a technical audience, the serious reader will find this a useful and interesting work.

———. 2006. Accelerating Universe: Observational Status and Theoretical Implications. Available at http://arxiv.org/PS_cache/astro-ph/pdf/0601/0601014.pdf (accessed September 1, 2006). The author has written a thoughtful and technical overview of the discovery of the accelerating universe and the various models of dark energy. Alternate interpretations and observational difficulties are recounted with honesty and detail, something often ignored in similar papers. The interested reader can easily skip the occasional mathematics in order to gain an understanding of the inherent difficulties in this area of research.

Randall, Lisa. 2005. *Warped Passages.* New York: HarperCollins. A popular-level survey of string and brane theories written by a leading researcher in the field of brane worlds. This thick volume will satisfy the reader who is interested in the nuts and bolts of branes but who lacks the mathematical background to read professional papers.

Rees, Martin. 2000. *New Perspectives in Astrophysical Cosmology,* 2nd ed. Cambridge: Cambridge University Press. This thin volume gives the popular-level reader a taste of some of the successes and remaining questions of twentieth century astronomy. Written by a radio astronomer involved in resolving the steady-state/big bang debate, his personal insights and sharp wit provide a unique perspective.

Rubin, Vera. 1997. *Bright Galaxies Dark Matters.* New York: Springer Verlag. In this series of essays and short scientific papers, Rubin surveys her considerable contributions to our understanding of the role of dark matter in galaxies. Equally important, her biographical writings paint a vivid picture of the discrimination she faced during the early years of her career as the rare woman in cosmology.

Sagan, Carl. 1980. *Cosmos.* New York: Random House. Based on his classic television series of the same name, Sagan demonstrates why he is the ultimate popularizer of science. The reader journeys with Sagan through the universe in search of the ultimate answers, including whether or not we are alone. In the process, the reader learns more about what it fundamentally means to be human. Although some of the science is dated, this text (and the revised video series recently released) are highly recommended to all those with an interest in cosmology.

———. 1996. *The Demon-Haunted World.* New York: Random House. Master popular science writer Sagan teaches the scientific method through engaging examples. More importantly, he lays out the difference between science and pseudoscience and explains the dangers the latter poses to a technological society.

Singh, Simon. 2004. *Big Bang.* New York: HarperCollins. In this charming popular-level volume, Singh takes the reader from ancient Greece to cutting-edge research in various aspects of cosmology. Scientific discoveries are placed in their proper context with historical facts not recounted elsewhere. Highly recommended for popular-level readers.

Smith, Sinclair. 1936. The Mass of the Virgo Cluster. *Astrophysical Journal* 83: 23–30. This is perhaps the first paper written in English to suggest that unseen material (dark matter) inhabits galaxy clusters. Although written for a technical audience, the introduction and conclusion are highly readable and of historical importance.

Spergel, D. N., R. Bean, O. Dore, et al. 2006. Wilkinson Microwave Anisotropy Probe (WMAP) Three Year Results: Implications for Cosmology. Available at http://arxiv.org/PS_cache/astro-ph/pdf/0603/0603449.pdf (accessed September 1, 2006). This technical paper is dense with observations from the WMAP cosmic microwave background probe and clearly demonstrates how these observations bolster the concordance model. The importance of these results made this preprint one of the most highly cited references in cosmology shortly after its release.

Sproul, Barbara. 1979. *Primal Myths*. San Francisco: HarperCollins. This work features an impressive compilation of creation myths from around the world. Although organized geographically, this work demonstrates the universality of certain aspects of creation myths, no matter the country of origin.

Starkman, Glenn D. and Dominik Schwarz. 2005. Is the Universe Out of Tune? *Scientific American* 293(2): 48–55. This popular-level article clearly explains the importance of acoustic waves in the early universe. Successes and lingering problems with the concordance model with regard to WMAP data are laid out without bias.

Susskind, Leonard. 2006. *The Cosmic Landscape*. New York: Little, Brown and Co. Susskind, a leading researcher in string theory, masterfully explains his speciality as well as topics such as brane theory and the anthropic principle with keen insight and considerable humor. This book is highly recommended for the reader interested in these topics but who lacks a technical background.

Tegmark, Max. 2004. Parallel Universes. In *Science and Ultimate Reality*, ed. John D. Barrow, Paul C. Davies and Charles L. Harper, Jr., 459–491. Cambridge: Cambridge University Press. One of the most detailed surveys of various multiverse models, where the definition of multiverse is taken in its broadest terms. Tegmark reviews the properties of each model as well as possible means of testing/observing each. The treatment is not mathematical but recommended for the serious rather than casual reader.

Trimble, Virginia. 1995. The 1920 Shapley-Curtis Discussion: Backgrounds, Issues, and Aftermath. *Publications of the Astronomical Society of the Pacific* 107: 1133–1144. Trimble offers a detailed review of the "Great Debate," including the personalities involved and the larger issues facing the society of the time. The points of contention are clearly laid out, giving each man his due in terms of where he was correct and where he was limited by the technology of the day. It is highly recommended for readers interested in early twentieth century cosmology.

Van den Bergh, Sidney. 1988. Novae, Supernovae, and the Island Universe Hypothesis. *Publications of the Astronomical Society of the Pacific* 100: 8–17. An excellent scientific and historical overview of the original confusion between novae and supernovae and the effect this had on the island universe debate. This work is recommended for readers interested in this important era in the history of modern cosmology.

————. 2000. *The Galaxies of the Local Group.* Cambridge: Cambridge University Press. This text is a detailed study of the galaxies in our cosmic neighborhood. Dense with facts and references, this volume will be of interest to readers curious about the vital statistics of our Local Group and its citizens.

Von Hippel, Ted, Chris Simpson, and Nadine Manset, eds. 2001. *Astrophysical Ages and Time Scales.* San Francisco: Astronomical Society of the Pacific. These conference proceedings cover a wide variety of techniques, successes, and problems in determining the ages of various astronomical objects and the universe at large. Most of the articles are written for a technical audience, but the serious reader will gain at least a flavor of the successes and challenges of these difficult areas of research.

Waller, William H. and Paul H. Hodge. 2003. *Galaxies and the Cosmic Frontier.* Cambridge, MA: Harvard University Press. This work is a detailed yet accessible survey of current understanding of galaxies and large-scale structure written by researchers in the field. It is recommended for the reader especially interested in these aspects of cosmology.

Webb, Stephen. 1999. *Measuring the Universe.* London: Springer. Written for the undergraduate astronomy major, the mathematics can easily be ignored by the interested popular-level reader. It is an excellent survey of the various rungs of the distance ladder and the techniques utilized at each level.

Wesson, Paul S., K. Valle, and R. Stabell. 1987. The Extragalactic Background Light and a Definitive Resolution of Olbers's Paradox. *Astrophysical Journal* 37: 601–606. The authors offer a definitive explanation of the solution of Olbers' paradox. Although appearing in a technical journal, the introduction and conclusion are accessible to the interested reader.

Wiita, Paul J. 2006. Active Galactic Nuclei: Unification, Blazar Variability and the Radio Galaxy/Cosmology Interface. Available at http://arxiv.org/PS_cache/astro-ph/pdf/0603/0603728.pdf (accessed September 1, 2006). Wiita has written a concise yet detailed survey of the various types of active galactic nuclei and the successes and limitations of the unified model. Written for a technical audience, the first half of the article is accessible to the interested reader, while the details in the latter half make it difficult going.

Yulsman, Tom. 2003. *Origins.* Bristol, UK: Institute of Physics. A journey through modern cosmology as seen through the eyes of key researchers, this popular-level book balances observation and theory, establishment and speculation.

INDEX

About the Author

KRISTINE M. LARSEN is Professor of Astronomy and Physics at Central Connecticut State University. She is the author of *Stephen Hawking: A Biography* (Greenwood, 2005).